# 104
# Structure and Bonding

Managing Editor:
D.M.P. Mingos

Springer
*Berlin*
*Heidelberg*
*New York*
*Hong Kong*
*London*
*Milan*
*Paris*
*Tokyo*

# Group 13 Chemistry II
## Biological Aspects of Aluminum

**Volume Editors:**
**H.W. Roesky and D.A. Atwood**

With contributions by
S. Anitha, D.A. Atwood, K. Berend, D. Bohrer,
B. Conley, V.M. Morsch, S. Oshiro, J. Patočka,
K.S.J. Rao, M.R.C. Schetinger, A. Strunecká,
G.B. van der Voet, F.A. de Wolff

Springer

The series *Structure and Bonding* publishes critical reviews on topics of research concerned with chemical structure and bonding. The scope of the series spans the entire Periodic Table. It focuses attention on new and developing areas of modern structural and theoretical chemistry such as nanostructures, molecular electronics, designed molecular solids, surfaces, metal clusters and supramolecular structures. Physical and spectroscopic techniques used to determine, examine and model structures fall within the purview of Structure and Bonding to the extent that the focus is on the scientific results obtained and not on specialist information concerning the techniques themselves. Issues associated with the development of bonding models and generalizations that illuminate the reactivity pathways and rates of chemical processes are also relevant.

As a rule, contributions are specially commissioned. The editors and publishers will, however, always be pleased to receive suggestions and supplementary information. Papers are accepted for *Structure and Bonding* in English.

In references *Structure and Bonding* is abbreviated *Struct Bond* and is cited as a journal.

Springer WWW home page: http://www.springer.de
Visit the SB home page at http://link.springer.de/series/sb/ or
http://link.springer-ny.com/series/sb/

ISSN 0081-5993
ISBN 3-540-43807-6
Springer-Verlag Berlin Heidelberg New York

CIP Data applied for



Springer-Verlag Berlin Heidelberg New York a member of BertelsmannSpringer Science + Business Media GmbH
http://www.springer.de

Printed in Germany



Typesetting: Scientific Publishing Services (P) Ltd, Madras
Production editor: Christiane Messerschmidt, Rheinau
Cover: Medio Technologies AG, Berlin
Printed on acid-free paper SPIN: 108 428 44 02/3020 – 5 4 3 2 1 0

## Managing Editor

Professor D. Michael P. Mingos

Principal
St. Edmund Hall
Oxford OX1 4AR, UK
*E-mail: michael.mingos@seh.ox.ac.uk*

## Volume Editors

Professor Herbert W. Roesky

Institut für Anorganische Chemie
der Universität Göttingen
Tammannstr. 4
37077 Göttingen, Germany
*E-mail: hroesky@gwdg.de*

Professor Dr. David A. Atwood

Department of Chemistry
University of Kentucky
Lexington, Kentucky, 40506 0055, USA
*E-mail: Datwood@pop.uky.edu*

## Editorial Board

## Structure and Bonding
## Also Available Electronically

For all customers with a standing order for Structure and Bonding we offer the electronic form via LINK free of charge. Please contact your librarian who can receive a password for free access to the full articles by registration at:

http://link.springer.de/orders/index.htm

If you do not have a standing order you can nevertheless browse through the table of contents of the volumes and the abstracts of each article at:

http://link.springer.de/series/sb/
http://link.springer-ny.com/series/sb/

There you will also find information about the

- Editorial Board
- Aims and Scope
- Instructions for Authors

# Preface

Aluminum, bound almost exclusively to oxygen in various combinations, is the most abundant metal in the earth's crust and, therefore, of great commercial potential. Once methods were developed (in the 1880's) to free useable quantities of the element from oxygen, applications for the element began developing rapidly. This growth has resulted in the ubiquity of the metal in today's world. Therefore it can be found intentionally introduced in many products in direct contact with human beings. It is commonly known that soluble forms of aluminum are toxic to living organisms. However, aluminum is not known to be bioavailable under everyday conditions. In fact, the solubility product of common aluminum compounds, such as $Al(OH)_3$ is so low as to make it essentially unavailable.

This volume of Structure and Bonding seeks to provide in one source, a resource where the basic science related to aluminum toxicity may be obtained. It should be stressed that this volume is not intended to be a warning to avoid contact with aluminum. Living organisms have adequate defenses to prevent aluminum toxicity under normal conditions. Rather the volume was created to simply provide an understanding of the biological effects of aluminum. As such, the present volume should be considered in the context of the companion volumes in this three-part series of Structure and Bonding. The first volume was devoted to fundamental developments in group 13 chemistry. The present volume is the second in the series and is meant to also address fundamental science, but more focused on biological aspects of aluminum. The third volume will focus on the applied group 13 chemistry.

Two types of publications are presented herein. The first set outlines the toxic effects of aluminum compounds on various living systems. The second set, comprised of two papers, deals with the formation and activity of aluminum fluoride compounds. The Volume begins with a chapter by Berend "Acute Aluminum Intoxication" that outlines the myriad toxic effects aluminum can have once it has by-passed an organisms' protective barriers. This occurs in humans, for example, when aluminum salts are used in medicine (a practice that has now been eradicated). The in-depth coverage of this topic provides an excellent background for understanding the chemical interactions associated with aluminum that are described subsequently in Chapters 2–4.

Chapters 5 and 6 signal the growing importance of combinations between aluminum and fluorine in living systems. Fluorine forms aluminum bonds

(583 kJ-mol$^{-1}$) that are perhaps bested only by scandium in strength (~590 kJ-mol$^{-1}$). Thus, anytime these two elements are present there is a high likelihood that they will unite. Strunecka explores in Chapter 5, "Aluminum Fluoride Complexes" the possibility that low-levels of fluoride, in blood plasma, and elsewhere, can have a synergistic effect with aluminum. This may occur through the $AlF_x$ compounds behaving as non-hydrolyzing phosphate analogues. Atwood explores this possibility further, from a chemistry perspective, in Chapter 6, "Fluoroaluminate Chemistry". In this chapter the formation and stability of various Al-F compounds is discussed. This will help elucidate the exact nature of the fluoroaluminate having the biological effects noted in Strunecka's studies.

With these six chapters this volume will provide the scientific researcher and non-specialist alike with a complete and up-to-date coverage of biological aluminum chemistry. The volume can also be used as a starting point to begin more detailed studies in this area. All of the relevant literature can be found in these chapters. Finally, we have to mention that we are chemists and therefore we have not evaluated completely the medical and toxicological implications of the chapters.

July 2002

H.W. Roesky
D.A. Atwood

# Contents

## Contents of Volume 103

## Group 13 Chemistry I
## Fundamental New Developments

Volume Editors: H.W. Roesky, D.A. Atwood



## Contents of Volume 105 (preliminary)

## Group 13 Chemistry III
## Industrial and Applied

Volume Editors: H.W. Roesky, D.A. Atwood

# Acute Aluminum Intoxication

Kenrick Berend[1], Gijsbert B. van der Voet[2], Frederik A. de Wolff[2]

[1] Diatel Curaçao, Jan Noorduynweg 81, Curaçao, Netherlands Antilles
*E-mail: kenber@attglobal.net*

[2] Toxicology Laboratory, Leiden University Medical Center L1-P, P.O. Box 9600, 2300 RC Leiden, The Netherlands

Despite the abundance of aluminum (Al) in nature, it has no known biological function in humans. On the contrary, wide ranges of toxic effects of Al to hundreds of cellular processes both in man and animals have been demonstrated in plants and aquatic animals in nature, experimental animals by several routes of exposure, and under different clinical conditions in humans. Manifestations of Al toxicity that are encountered vary substantially, with major differences arising due to different sources and varying intensity of the Al burden in combination with differences in individual susceptibility. Al neurotoxicity occurs only under extreme conditions and the conditions are the worst in acute Al encephalopathy where myoclonic jerks, convulsions and a high mortality rate are the major manifestations. Especially hemodialysis patients are at risk when the dialysate is contaminated with high Al concentration, but also the use of Al in bladder irrigation and the use of Al containing cement in otosurgery has been related to acute Al encephalopathy. Therefore, every effort should be undertaken to limit the use of compounds containing Al in medicine.

**Keywords:** Aluminum, Intoxication, Acute, Hemodialysis



Structure and Bonding, Vol. 104

## 1 Introduction

Aluminum (Al) is present in small amounts in mammalian tissues but it has no recognized physiological role. On the contrary, wide ranges of toxic effects of Al have been demonstrated in plants and aquatic animals, in experimental animals by several routes of exposure, and under different clinical conditions in humans. Its neurotoxic effect on living organisms is beyond any doubt and Al has been shown to interfere with a variety of cellular metabolic processes in the nervous as well as several other systems. About eight percent of the earth's crust consists of Al (w/w) and, despite its abundance, its toxicity remained disputed until the discovery of Al-related diseases in renal patients. This is due to the fact that the bioavailability of Al is very low after oral ingestion and even lower after inhalation [1–4]. These protection barriers are bypassed during hemodialysis and during this treatment, trace elements can cross the artificial kidney and enter the blood of the patients. Due to the fact that patients are exposed to some 400 liters of water on a weekly basis, patients can be exposed to huge amounts of the toxic element and evidence of Al as a neurotoxic trace metal was established in the early 1970s after many years of uncertainty. Al neurotoxicity was initially described in 1886 [5] and rediscovered in 1921 [7] and 1937 [6]. Animal studies showed that local administration or application of Al to the brain caused animals to develop a seizure disorder [6]. In 1921 a possible single case of industrial poisoning with Al was reported [7] but it lasted until 1962 before the second case with industrial-related Al encephalopathy was described. In this case speech difficulties, seizures and pulmonary fibrosis was associated with the inhalation of Al-containing dust in an Al plant [8] and in 1975 a similar case of neurotoxicity was described in association with increased brain Al concentration [9]. These were the first reports to suggest that Al neurotoxicity could occur in humans. In 1972 a new distinct neurological disease, dialysis encephalopathy, dialysis dementia, or chronic Al encephalopathy with speech disturbances, personality changes, seizures and myoclonus was described in dialysis patients [10] and numerous similar reports followed thereafter [11–46]. Subsequently, it was found that certain species of animals, cats and rabbits developed neurofibrillary tangles following exposure to Al [47–56], which prompted the suggestion that Al was a possible cause of Alzheimer's disease. However, this theory was disputed because not all patients with Alzheimer's have high brain levels of Al and the senile plaques that are common in Alzheimer's disease are not seen in experimental Al toxicity [57]. Another reason for disregarding a role for Al in Alzheimer's disease has been the fact that the incidences of cognitive impairment and Alzheimer's disease are not increased in renal patients with high s-Al levels. Nevertheless, these patients cannot be compared with the general population, because dialysis patients have a reduced longevity and may be protected by other compounds like silicon [58–61].

It remains remarkable that, after the discovery of chronic Al encephalopathy and numerous reports of patients suffering from this syndrome, fourteen years passed before the first publication on acute Al neurotoxicity appeared [62]. Acute Al encephalopathy is a devastating, often fatal disease that is the result of iatrogenic exposure to Al. Unfortunately, in spite of efforts to avoid Al exposure as much as possible, recently new sources of serious exposure have been added to the growing list of sources of Al exposure. In contrast to a fascinating history and abundant literature, many uncertainties about Al toxicity still exist. Clinical data in humans on acute Al neurotoxicity are very limited and we will, therefore, compare our experience in one of the two documented outbreaks in dialysis centers, with that of the literature.

## 2 Case Report

### 2.1 Course of Events and Clinical Symptomatology

In Curacao, the major island of the Netherlands Antilles with a population of 130,000 inhabitants, distilled seawater from the water plant was used without further purification for hemodialysis for several decades. Unfortunately, two months before the planned installation of a water treatment system including a reversed osmosis (RO) in the dialysis center Diatel, a new distribution pipe supplying water to a dialysis center on the island was installed in 1996. To protect it from corrosion, this pipe was lined on the inside with a cement mortar. Because of the aggressiveness of the distilled water, calcium and Al leached from the cement mortar into the water used to prepare dialysate. At the time of replacement of the new conduit pipe, 29 patients were dialyzed in the dialysis unit. Patients were dialyzed three times per week during 3.5 to 4.5 hours using hollow fiber kidneys (Fresenius, F6 and F8). Untreated tap water was used to manufacture dialysate after passage through three particulate filters (25, 10, and 5 mm, respectively). The intended Ca content of the dialysate was 1.75 mmol/L. None of the patients used Al-containing phosphate binders. After the installation of the new water distribution pipe, the patients were exposed to the contaminated water from May 21, 1996, until June 29, 1996. This caused a possible hard water syndrome and definite acute Al intoxication in 27 patients and led to the death of 10 patients.

In the third week of June 1996, 25 of 27 patients had minor symptoms of nausea, vomiting and post-dialysis hypercalcemia. On June 20, 1996, the Ca-carbonate concentration of tap water delivered to the dialysis unit was 43 mg/L (17.2 mg/L or 0.45 mmol/L as $Ca^{2+}$), which is above the standard of 2 mg/L advised by the Association for the Advancement of Medical Instrumentation [63]. In the past, the Ca-carbonate concentration had been 15 to 20 mg/L (6.0 to 8.0 mg/L or 0.16 to 0.21 mmol/L $Ca^{2+}$). At that time the diagnosis "hard water syndrome" [64] was made, due to leaching of calcium from the water distribution pipe. The use of Ca-carbonate and vitamin D preparations was stopped and although there was initial amelioration of the symptoms in almost all patients, the improvement was incomplete whereafter the dialysis unit was closed on June 30, 1996. All patients were referred to the hospital for dialysis where the symptoms of nausea, vomiting, as well as the hypercalcemia disappeared after a single dialysis with low Ca dialysate (1.50 mmol/L). After installation of an RO system, including a deionizer, dialysis was resumed at the Diatel unit on July 10, 1996. Although the patients were dialyzed on ultra pure water, either in Diatel after the installation of the RO on July 10, 1996, or in the local hospital, before the manifestation of serious symptoms, after a lag time of several days to weeks several patients had severe neurotoxic symptoms. Eight patients (Table 1, patients 2 through 5 and 7 through 10) had to be admitted to the hospital because of severe neurological symptoms (disorientation, myoclonus, convulsions and coma, n = 7) or unexplained sepsis (n = 2) with a delay of days to three weeks after the last dialysis with contaminated water (Fig. 1). Attempts to treat myoclonic jerks or seizures with intravenous phenytoin or diazepam were unsuccessful. When the diagnosis of "acute Al intoxication" was established on July 25, 1996, nine patients had died. The 10th patient, who was comatose, was transferred to a hospital in Florida (Florida Hospital, Orlando, FL, USA). Daily high-flux dialysis treatment combined with desferrioxamine administration was unsuccessful, and he died on August 24, 1996. Of the 27 patients who had a similar exposure (~60 hours) to the contaminated dialysate, 10 died from acute Al encephalopathy, whereas 17 patients had no or only minor symptoms and survived. The first Al tests in the water supply were performed the first week of July 1996, and values of 550 to 690 μg/L were observed. The water company flushed the pipe several times, but because the values remained above 500 μg/L, 1000 m of the initial 2200 m pipe was bypassed. Nevertheless, although the Al level at the water plant was below 5 μg/L, the Al levels of the tapwater in Diatel remained high (up to 443 μg/L several months after the bypass), and above 100 μg/L for more than two years [18].

Of the 29 patients, two patients who had been dialyzed for less than a week (one at the end of May, and one at the end of June 1996) on the contaminated

**Table 1.** Characteristics of non-survivors and survivors

| Patient no. | Age (years) | [Al]s (μg/mL) | Sex | Weight (kg) | Exposure (h) | Diuresis (mL/day) | PTH (pg/mL)mL) | Albumin (g/L) | Δt[e] E[f] | Δt[e] D[g] | Treatment (weeks) |
|---|---|---|---|---|---|---|---|---|---|---|---|
| Non-survivors (n = 10) | | | | | | | | | | | |
| 1 | 61 | – | f | 66 | 52 | 0 | 28 | 35 | 4 | 9 | – |
| 2 | 79 | – | f | 33 | 51 | 0 | 105 | 32 | 7 | 12 | – |
| 3 | 54 | 1189 | m | 85.5 | 68 | 150 | 52 | 34 | 2 | 12 | – |
| 4 | 64 | 725 | f | 52.5 | 52 | 0 | 22 | 35 | 15 | 18 | – |
| 5[a] | 51 | 894 | f | 40 | 68 | 0 | 72 | 30 | 7 | 20 | – |
| 6 | 69 | – | f | 39.5 | 39 | 2000 | – | 28 | 7 | 22 | – |
| 7[a] | 62 | 1275 | m | 62 | 60 | 0 | 8 | 34 | 5 | 23 | – |
| 8[a] | 63 | 359 | f | 47 | 68 | 0 | 325 | 36 | 14 | 23 | – |
| 9[a] | 74 | 517 | f | 86 | 64 | 0 | 32 | 32 | 21 | 24 | – |
| 10 | 65 | 696 | m | 64 | 68 | 0 | 480 | 35 | 21 | 30 | – |
| Mean | 64 | 808 | – | 57.5 | 59 | – | 125 | 33 | | | |
| SEM | 3 | 127 | – | 5.9 | 3 | – | 55 | 1 | | | |
| Survivors (n = 17) | | | | | | | | | | | |
| 11 | 64 | 490 | m | 57 | 68 | 0 | 710 | 40 | – | – | 75 |
| 12 | 57 | 395 | m | 65 | 59.5 | 0 | 80 | 39 | – | – | 91 |
| 13 | 50 | 321 | f | 87 | 68 | 0 | 21 | 34 | – | – | 16 |
| 14 | 44 | 319 | m | 97 | 62 | 0 | 630 | 39 | – | – | 13 |

| | | | | | | | | | | | |
|---|---|---|---|---|---|---|---|---|---|---|---|
| 15 | 49 | 318 | m | 57 | 66 | 200 | 50 | 36 | – | – | 24 |
| 16 | 23 | 315 | f | 104 | 68 | 0 | 97 | 39 | – | – | 8 |
| 17 | 66 | 303 | f | 78.5 | 64 | 60 | 280 | 34 | – | – | 18 |
| 18 | 57 | 301 | f | 90 | 68 | 350 | 170 | 37 | – | – | 5 |
| 19 | 50 | 248 | f | 57.5 | 64 | 1000 | 65 | 36 | – | – | 13 |
| 20 | 56 | 207 | f | 103 | 68 | 0 | 82 | 36 | – | – | 16 |
| 21 | 50 | 205 | m | 104 | 60 | 2 | 700 | 37 | – | – | 82 |
| 22 | 57 | 187 | f | 101 | 64 | 50 | 1200 | 32 | – | – | 16 |
| 23 | 57 | 182 | f | 95 | 62.5 | 1100 | 100 | 30 | – | – | 9 |
| 24 | 41 | 166 | m | 91.5 | 68 | 0 | 3 | – | – | – | 18 |
| 25 | 60 | 132 | m | 97.5 | 67 | 1200 | 270 | 39 | – | – | 4 |
| 26 | 41 | 116 | m | 93.2 | 60 | 200 | 680 | 38 | – | – | 11 |
| 27 | 56 | 113 | f | 91.5 | 67 | 1400 | 99 | 33 | – | – | 4 |
| Mean | 52[c] | 255[c] | – | 86.5[c] | 65 | – | 308 | 36[d] | | | |
| SEM | 2 | 25 | – | 4.1 | 1 | – | 84 | 1 | | | |

[a] Patients in whom post-mortem examination was performed.
[b] [Al]s: serum aluminum concentration.
[c] $P < 0.01$.
[d] $P < 0.05$, survivors vs. non-survivors.
[e] rt E/D pt: delay in time in days after end of exposure to AL, the onset of encephalopathy (E[f]), and death (D[g]).
with permission Blackwell Sciences [17]. Adapted.

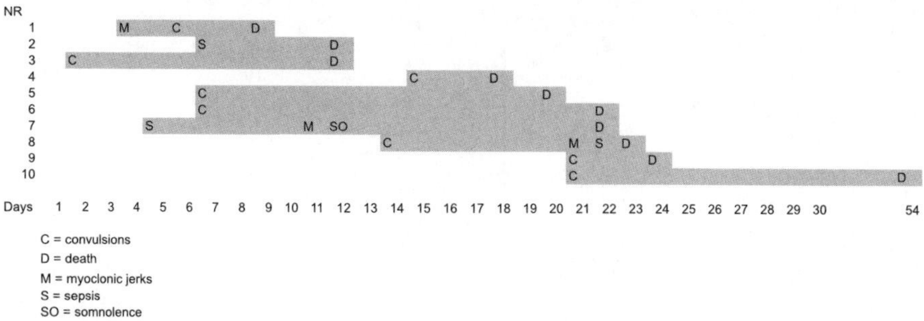


**Fig. 1.** Days between the onset of severe symptoms after the end of exposure to aluminum-contaminated dialysate

water, had s-Al levels of 16 μg/L and 50 μg/L (<60 μg/L normally seen in dialysis patients without symptoms) and were therefore excluded from further evaluation. Table 1 shows the characteristics of non-survivors and survivors. It is not certain if patient No. 6 did or did not develop acute Al encephalopathy, because of the 27 patients she probably had the lowest Al burden. The patient with a diabetic nephropathy had been on hemodialysis for only two months and had a considerable residual kidney function with normal diuresis. Compared with the other patients she was exposed for the shortest period (39 h vs. 59–65 h) because she was admitted twice to the hospital for a bleeding duodenal ulcer and hypercalcemia. Hypercalcemia was due to a combination of factors consisting of the use of a calcium phosphate binder, vitamin D, a higher than usual calcium content of the water supply, the availability of calcium in the dialysate and the Al contamination. The calcium level normalized after she had been dialyzed in the hospital with a low-calcium dialysate. The patient nevertheless developed convulsions and twitching of her hands that might have been a sign of Al neurotoxicity, especially because all EEG, performed two weeks before the start of convulsions showed, in retrospect, signs of Al encephalopathy (Table 2). These symptoms, however, could also be due to a metabolic or uremic disorder because she had skipped some dialysis sessions. These twitchings were also absent in the other patients. After a one week of hospital admission dialysis treatment was stopped because of worsening of the general condition with anorexia and she died 22 days after the end of exposure to the contaminated dialysate (Fig. 1). No blood samples had been available for Al testing.

## 2.2 EEG

Table 2 shows the EEG recordings before treatment in 12 patients (patient Nos. 1, 3, 4, 6, 7, 10–13, 20, 23, 25) and five patients after treatment (patient Nos. 11, 12, 15, 20, 25). The abnormalities in the EEG recordings of the patients who died were the most severe and consisted of an abnormal ground pattern, some periods of slower high voltage activity, bilateral synchronous with some frontal intermittent recurrent delta activity, sometimes followed by short periods of suppression. After termination of desferrioxamine therapy an

EEG was performed in five patients (patient Nos. 11, 12, 15, 20 and 25). Minor non-specific signs possibly related to Al encephalopathy were found after therapy in two patients (Nos. 11 and 25). Slightly abnormal EEG became normal in two patients (patient Nos. 12 and 20). In patient No. 15 with a normal EEG after therapy no EEG had been obtained before therapy.

## 2.3 Pathology (Table 3)

Post-mortem examination was done in four of the 10 patients (Patient Nos. 5, 7–9). Brain Al contents of cerebral cortex of these four patients were 0.93 to 1.81 μg/g dry weight (normal 0.14–0.22 μg/g; values in patients with chronic Al encephalopathy 0.4–3 μg/g) [65]. On light-microscopic evaluation of the gyrus frontalis superior and inferior, hippocampus, nucleus lentiformis and insula, capsula interna and caudatus, thalamus, nucleus dentatus cerebelli, brainstem, and pons, non-specific changes of some fibrohyalin thickening of the wall of the blood vessels and brain edema were noted.

In patient No. 9, additional Al staining was performed and positive coloring was found centrally in calcifying areas in the plexus chorioidea of the hippocampus and in the wall of the vessels with mineralization in the globus pallidus. No Al staining was found in the epithelium of the plexus choreoidea, corpora amylacea, endothelium, neurons or glia cells. In Bodian silver staining scattered neurofibrillary tangles with tau protein were found in the hippocampus, similar to other studies [47–51, 53, 55–57, 66]. Iron (Perls) and calcium (Alizarine) staining showed the same pattern of staining as in Al.

## 2.4 Predisposing Factors

When the ten non-survivors are compared as a group, they were older (64 ± 3 years vs. 52 ± 2 years, $P < 0.01$), had a lower body weight (57.5 ± 5.9 kg vs. 86.5 ± 4.1 kg, $P < 0.01$) and lower serum albumin concentrations (33 ± 1 vs. 36 ± 1 g/L, $P < 0.01$) as a parameter for the feeding state of the patients. Anuria tended to be more common in the non-survivors (8 out of 10 vs. 8 out of 17, $P > 0.05$). S-Al concentrations, available in seven non-survivors, were significantly higher than in the survivors (808 ± 127 vs. 255 ± 25 mg/L, $P < 0.01$).

The 17 patients who survived the dialysis with contaminated dialysate were significantly younger than those who died (Table 1) and the average body weight in the non-survivors was almost 30 kg below that of the survivors [17]. Diabetes mellitus was more frequent in the non-survivors than in the survivors (60 vs. 16%, $P = 0.11$), as was the proportion being anuric (80 vs. 47%, $P = 0.12$). Survivors tended to have higher serum PTH concentrations (survivors, 308 ± 84 pg/mL; non-survivors, 125 ± 55 pg/mL, normal, 11 to 62 pg/mL). Shortly after the intoxication, the mean corpuscular volume in the patients was low (70.25 ± 6.22, normal 81–99 fL) in seventeen patients.

**Table 2.** EEG before and after therapy, serum aluminum levels and duration of treatment[a]

| No.[b] | s-Al | R/weeks | Initial EEG | EEG survivors, after therapy |
|---|---|---|---|---|
| 1 | – | – | Abnormal ground pattern. Delta activity with a maximum bilateral frontoparietal and also bilateral synchronous spikes and triphasic waves | |
| 2 | – | – | ND[c] | |
| 3 | 1189 | – | Abnormal ground pattern. More or less continuous slow to very slow peak waves | |
| 4 | 725 | – | Normal ground pattern. Left central sharp theta waves (middle high) | |
| 5 | 894 | – | ND | |
| 6 | – | – | Some periods of slower high voltage activity, bilateral synchronous with some frontal intermittent recurrent delta activity | |
| 7 | 1275 | – | Abnormal ground pattern. Diffuse abnormal EEG | |
| 8 | 359 | – | ND | |
| 9 | 517 | – | ND | |
| 10 | 696 | – | Abnormal ground pattern. More or less continuous high voltage delta waves followed by short periods of suppression | |
| 11 | 490 | 75 | Multifocal theta, bilateral frontotemporal, followed by asymmetrisch waves with a mitigated epileptiform aspect | Left temporal slow theta and delta waves of middle high to high |
| 12 | 395 | 91 | Minor abnormality. Intermittent right temporal some theta activity | Normal |

| | | | | |
|---|---|---|---|---|
| 13 | 321 | 16 | Some signs of aluminum encephalopathy | ND |
| 14 | 319 | 13 | ND | ND |
| 15 | 318 | 24 | ND | Normal |
| 16 | 315 | 8 | ND | ND |
| 17 | 303 | 18 | ND | ND |
| 18 | 301 | 5 | ND | ND |
| 19 | 248 | 13 | ND | ND |
| 20 | 207 | 16 | Scattered some delta/theta activity temporobasal and dorsum of the brain | Normal |
| 21 | 205 | 82 | ND | ND |
| 22 | 187 | 16 | ND | ND |
| 23 | 182 | 9 | Diffuse hypofunctional abnormal. Not typical of aluminum encephalopathy | ND |
| 24 | 166 | 18 | ND | ND |
| 25 | 132 | 4 | Scattered multifocal, frontotemporal irregular and irritative | Left frontoparietal slow waves |
| 26 | 116 | 11 | ND | ND |
| 27 | 113 | 4 | ND | ND |

[a] In all patients who had a EEG recording had inconsistent signs of a metabolic encephalopathy. The abnormalities in the EEG recordings of the patients who died were the most severe and consisted of an abnormal ground pattern, some periods of slower high voltage activity, bilateral synchronous with some frontal intermittent recurrent delta activity, sometimes followed by short periods of suppression. After termination of desferrioxamine therapy an EEG was performed in five patients (patient Nos. 11, 12, 15, 20 and 25). Minor non-specific signs possibly related to aluminum encephalopathy were found after therapy in two patients (Nos. 11 and 25). Slightly abnormal EEG became normal in two patients (patient Nos. 12 and 20). In patient No. 15 with a normal EEG after therapy no EEG had been obtained before therapy;

[b] Patient number, see Table 1.

[c] ND not done.

**Table 3.** Post-mortem Al in serum, brain, bone and liver in patients Nos. 5, 7–9[a]

| Patient no. | Normal values | Chronic Al encephalopathy in dialysis patients | 5 | 7 | 8 | 9 |
|---|---|---|---|---|---|---|
| **Serum** Al µg/l | <10 µg/L [23, 129] (40–60 µg/L: dialysis patients) [23, 129] | >100 µg/L [20] | 894 | 1275 | 696 | 517 |
| **Brain** | | | | | | |
| Hippocampus | <0.4 µg/g dry weight [65] | 0.4–3.7 [65] | NA[b] | 1.14 | 1.3 | 1.03 |
| Gyrus temporalis inferior | <0.4 µg/g dry weight [65] | 0.4–3.7 [65] | 1.75 | 1.04 | 1.4 | 0.93 |
| Gyrus frontalis superior | <0.4 µg/g dry weight [65] | 0.4–3.7 [65] | 1.81 | 1.19 | 1.51 | 1.3 |
| Liver | <2 µg/g [129] | >30 µg/g [42] | 4.7 | 51.7 | 32.7 | 43 |
| **Bone** | <2 µg/g [23, 129] | 50.8 + 29.5 µg/g [197] | 77 | 7.54 | 88.7 | 21.9 |

[a] Al levels in serum, liver and bone are much higher than in the brain, which shows the effectiveness of the blood brain barrier and the storage capacity of liver and bones. Nevertheless, severe neurological symptoms are due to an extremely narrow margin of safety between normal and toxic levels of Al in brain tissue and the lethal concentrations exceed the normal level by only a factor 8–10 in literature studies [31] and a factor 3–4 in this study. Epilepsy and myoclonic jerks are prominent signs of neurological dysfunction due to Al encephalopathy . The gray matter of the brain is vulnerable to Al-induced pathologic changes and dysfunction of the hippocampus often manifests as epilepsy. Myoclonus is often associated with abnormally increased excitability of cortical structures [274]. The gyrus temporalis inferior and gyrus frontalis superior are related to speech disorders and dyscalculia as can be seen in chronic Al encephalopathy and Alzheimer's disease [275].

[b] NA not available.

## 2.5 Detoxification Treatment and Follow-Up of the Survivors

The 17 surviving patients had elevated s-Al concentrations but were fairly asymptomatic and had to be treated with intravenous desferrioxamine (DFO) and high-flux dialysis according to the treatment schedule of Table 4. For logistic reasons, especially the inability to provide high-flux dialysis to all patients on short notice, these patients were transferred to six different hospitals in the Netherlands. In the initial treatment week no serious side effects were observed. In the second week one patient had a short period with myoclonic jerks during dialysis. In the weeks following the treatment only minor neurological or ophthalmologic symptoms were observed, expressed by slight headache (n = 3), and blurred vision (n = 1). None of the patients had serious side effects of the treatment with desferrioxamine. One patient had an allergic reaction expressed by generalized itching during infusion of DFO, but when DFO was infused the next session after steroid infusion, no symptoms were recorded. One patient had painful localized swellings on both legs, diagnosed as erythema nodosum. This is not a known side effect of desferrioxamine. The lesions disappeared rapidly with steroid treatment. One patient had an infection of a dialysis shunt (Goretex graft).

A large variation of the duration of the detoxification was noted (Table 1). Only five patients could be treated for less than 10 weeks, while twelve other patients had to be treated for up to 91 weeks. The treatment was discontinued in patients in whom both s-Al and the increment of s-Al after desferrioxamine treatment were below 50 µg/L at two successive occasions). The treatment duration was significantly related to the residual diuresis as all patients with a residual diuresis of a liter/day or more could be treated for less than two months (Fig. 2). Other studies have also established the protective capacity of an even minimal functioning kidney [30, 67]. None of the patients died during treatment with desferrioxamine, and six patients (patient Nos. 11–13, 22, 25, 26) died more than one year after termination of the desferrioxamine treatment, due to causes unrelated to the Al intoxication. As of November 2001, more than five years after the intoxication episode, 12 of the 17 surviving patients (patient Nos. 11, 12, 14, 16–24) are still alive and none of the patients developed any clinical signs of Al toxicity, like speech disturbances, cognitive defects, bone fractures or dementia-like symptoms.

## 2.6 Discussion

A unique set of circumstances was responsible for symptoms resembling hard water syndrome [64] followed by an epidemic of acute Al encephalopathy in a dialysis unit (Diatel) on the island of Curaçao. A tragic coincidence is that the intoxication happened about two months before the planned installation of a water treatment system with deionization and reverse osmosis (RO). Traditionally, municipal water had been used for more than two decades without extended purification for the production of dialysate. The pure

**Table 4.** Practical approach of severe aluminum intoxication in hemodialysis patients, using Monday as the starting day

| | Aluminum levels >300 μg/L, or post desferrioxamine administration >300 μg/L. Infusion time one hour. Four times a week dialysis | | Aluminum levels <300 μg/L, or post desferrioxamine administration <300 μg/L. Three times a week dialysis | |
|---|---|---|---|---|
| Therapy | Desferrioxamine administration 5 mg/kg, 5 hours before dialysis (1 h infusion time) | High flux dialysis of at least four hours F 80 or similar | Desferrioxamine administration 5 mg/kg at the last hour of dialysis (1 h infusion time) | High flux dialysis of at least four hours F 80 or similar |
| Day 1 Monday | Yes | Yes, four hours after termination of infusion | Yes | Yes |
| Tuesday | No | Yes, preferably 20 hours after termination of last dialysis session (within 72 hours) | No | No |
| Wednesday | No | No | Yes | Yes |
| Thursday | Yes | Yes, four hours after termination of infusion | No | |
| Friday | No | Yes, preferably 20 hours after termination of last dialysis session | No | Yes |
| Saturday | No | No | No | No |
| Sunday | No | No | No | No |

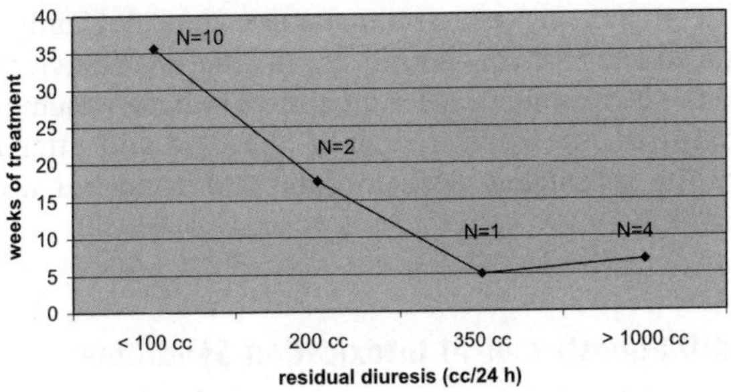


**Fig. 2.** Mean duration of desferrioxamine treatment (weeks) and residual diuresis in seventeen patients with aluminum intoxication

(Al < 5 μg/L), distilled drinking water, which is one of the purest drinking water sources worldwide [68], was in fact indirectly responsible for the fatal outcome. The water entering the mains is aggressive, soft and the alkalinity low (maximum Langelier index (LI) between −0.5 and −1.5, hardness as $CaCO_3$ is 15–20 mg/L, pH 8.5 to 9.5) [18]. The water utilities did not condition the water to make it less corrosive but installed other pipes with a "protective" cement coating [18]. Water distribution pipes are commonly coated on the inside with cement mortar to protect them from corrosion by the electrochemical action of water. Although cement-mortar linings have been successful in protecting iron pipes against electrochemical corrosive effects, recent research indicates that water can leach cementitious materials from this lining for up to four years. Adverse effects of these coatings on the water quality are generally an increase of calcium and pH [69], but recent reports did not mention the possibility of Al contamination into drinking water from cement linings [18]. The water distribution company had replaced the water distribution pipe to help the dialysis center with the low water pressure problem, but had been unaware of the fact that these pipes are inappropriate for the local water conditions. Conditions that enhance Al leaching include low water alkalinity and hardness, relatively high water temperatures, and low or intermittent water flow. Unfortunately, all of these factors coincided in this situation, especially because Diatel was located at the dead-end of a water main, resulting in the highest Al concentrations on the island. The Al level rose from about 16 μg/L to at least up to 690 μg/L. Clearly, no one, not even the water industry in particular, had ever been aware of the possibility of Al contamination during distribution, as this had never happened before in dialysis history [18].

## 2.7 Legal Consequences

The incident caused a flood of media coverage and a criminal investigation followed with the help of the local and Dutch Health Inspectorate. Thereafter, the prosecutor dismissed the case and held that the intoxication was caused by a unique and complex set of circumstances that not only involved the physicians, but also external consultants for the dialysis center, the local Water Authorities and the government. After an appeal of family members of some of the patients that died, the court reopened the case and after a preliminary judicial inquiry the physicians were initially convicted but acquitted at the Court of Appeal.

## 3 Symptoms and Diagnostics of Al Intoxication Syndromes

There are wide variations in the clinical manifestations and degrees of expression of toxicity as patients are exposed to Al. In renal disease Al accumulation in tissues and its resulting toxicity is mainly due to poor elimination, since the kidney normally excretes the metal. There is no simple relationship between the amount of Al to which the patient is exposed and the occurrence of the clinical manifestations of the disease; apparently there is considerable interindividual difference in sensibility since the symptoms developed in some patients but not in other equally exposed patients. Some of the variations in the clinical features in Al toxicity in the past most likely occurred because of differences in both the rapidity and magnitude of Al loading. Therefore, the manifestations of Al toxicity that are encountered vary substantially, with major differences arising due to different sources and varying intensity of the Al burden.

### 3.1 Definitions of Al Intoxications

In the past several clinical patterns have been described. The most important recognized clinical patterns or types of Al toxicity include two types of encephalopathy. Firstly, the "classical" dialysis dementia sometimes referred to as "dialysis encephalopathy syndrome (DES)" or dementia dialytica [24, 28, 37, 42, 70–74] and secondly, the "acute" or "subacute" Al encephalopathy [41]. There are also two types of bone disease – either osteomalacia with bone fractures and proximal myopathy or aplastic bone disease [41, 75, 76]. There is quite some confusion in the definitions of Al toxicity in the literature. Because there seems to be an obligatory lag phase of at least several days to weeks for symptoms to occur, "acute" Al encephalopathy, defined as a direct result of a single overdose, probably does not exist. Because of the long lag phase of several months to years necessary to develop the "chronic" dialysis encephalopathy and also because "acute" Al encephalopathy has an abrupt, sudden onset of symptoms one can understand why the term "acute" is used instead of the more descriptive "subacute". The descriptions "dialysis dementia" [37, 42, 46, 73, 74] and "dialysis encephalopathy" [33–36, 38, 40, 41, 78] are also unfortunate because true dementia is rare in Al encephalopathy [73] and non-dialyzed patients can also develop these symptoms [78]. There are also many dialysis-related encephalopathy syndromes unrelated to Al. As an example, the desequilibrium syndrome and sodium disturbances can also produce seizures [79]. In addition to this, some publications on "acute Al intoxication" [17, 80] refer to the predominant neurological symptoms, while other publications only described gastrointestinal symptoms [20, 81, 82].

To avoid this confusion it is necessary to separate acute Al intoxication into "acute Al gastroenteropathy" and "acute Al encephalopathy". Chronic Al gastroenteropathy as such probably does not exist, because patients will terminate the oral exposure because of complaints.

## 3.2 Acute Al Gasteroenteropathy

Gastroenterological symptoms due to Al toxicity have been described as massive gastrointestinal intoxication and as peritoneal dialysis fluid contamination with Al [81–83].

### 3.2.1 *Oral Exposure*

The potential negative effects of exposure to Al do not only reflect the dose, but also the percentage of the dose that is absorbed and retained. The effectiveness of the gut as a protective barrier is illustrated by the observation that only 0.04–1% of oral doses of Al appears to be retained [1–4, 59, 84–87]. However, very high concentrations of Al in drinking water may have health consequences. Drinking water plants commonly use Al sulfate as a flocculent. Unfortunately, human errors can increase human exposure to excessive amounts. On 6 July 1988, twenty metric tons of Al sulfate were accidentally emptied into the treated water reservoir at Lowermoor Water Treatment Works, that supplied the Camelford area of North Cornwall, UK. Twelve thousand residents and 8000 tourists were exposed for 3 days and when the error was detected corrective measures were started. The water tasted unpleasant because of its acidity and Al content up to 620 mg/L; it also contained abnormally high amounts of copper, lead and zinc dissolved from water pipes. Consumers noted an unpleasant taste and reported a variety of acute symptoms such as mouth ulceration, joint pain, gastrointestinal disturbance, skin rashes, fatigue, nausea and vomiting, itching and sore eyes [82, 88, 89]. Nevertheless, initial toxicological assessment was that thc limited exposure would not cause acute or chronic disease [82]. Two years later, however, of the 20,000 persons exposed to the contaminated water 400 people had illnesses that were attributed to the incident [88]. Bone Al concentrations remained elevated for 6 to 7 months, but were absent at follow-up 19 months after the accident in at least two affected individuals [88, 89]. In one study some patients demonstrated poor short-term memory and concentration disturbances which were regarded as a possible organic brain dysfunction due to the accident in one study [20]. Thus, Al absorption and deposition in the body can be significant when normal individuals are exposed to highly contaminated water, even when exposure is short. This may seem remarkable, considering the fact that Al-related bone disease in dialysis patients disappears after kidney transplantation and the rate of removal of Al from bone seems to be independent of graft function and bone turnover [90, 91]. One would expect that healthy kidneys should be able to eliminate Al in these circumstances. However, it has been demonstrated by special Al staining that Al-induced argyrophilic degradation products remain in the cellular cytoplasm in unchanged fashion up to 10 years after renal transplantation, when the renal Al excretion was normalized [92].

### 3.2.2
### *Peritoneal Membrane*

Another route of exposure resulting in gastrointestinal symptoms has been in peritoneal dialysis. In three of the centers supplied with contaminated peritoneal fluid in Paris, Glasgow and Edinburgh regular monitoring of the s-Al disclosed unexpected high Al levels. In 14 of 19 patients plasma Al levels exceeded 470 μg/L; four patients had symptoms of nausea, vomiting, cramp like abdominal pain and general malaise which appeared within three days of first using the contaminated fluid. The cause of the syndrome was recognized before irreparable harm was done, and all patients eventually recovered. Al concentrations in the affected batches ranged from 613–1440 μg/L. Within a month time, s-Al rose to peaks of 267–1440 μg/L [81, 83]. The mechanism of these symptoms is unknown; they seem to be due to a direct irritation of the bowel wall. We are unaware of any studies on neurotoxic damage to the autonomic nerves of the bowel due to AL.

## 3.3
## Clinical Symptoms of Acute Al Encephalopathy and Chronic Al Encephalopathy (also Referred to as Classical "Dialysis Dementia")

Slow, relatively "low exposure" accumulation of Al over a period of years can lead to a number of clinical manifestations, some of which seem to be "bypassed" in acute Al encephalopathy due to extremely high exposure to Al. Al encephalopathy is a clinical syndrome and, as can be seen in Table 5, there are similarities and differences in the neurological symptoms of acute and chronic Al encephalopathy. In chronic Al encephalopathy microcytic anemia [41, 93, 95–98] and EEG changes [99–104] can precede clinical symptoms [105]. It is unknown if these symptoms can also precede the clinical symptoms of acute encephalopathy. In contrast to acute Al encephalopathy, where speech disturbances are absent, speech disorders are an important presenting clinical sign of neurotoxicity in chronic Al encephalopathy. The neurological basis of the speech apraxia is obscure but it appears to have elements of dysarthria and dysphasia [33, 73]. The initial episodic nature of the condition suggests that it may be due to a transient toxic state as often occurs after dialysis with contaminated dialysate, but it might as well be part of a complex seizure state [33]. The neuropil spongiosis of the upper layers of the temporal cortex was thought to be relevant to the language disorder [103]. Besides communication disturbances, cognition disturbances and fracturing osteomalacia and/or myopathy can also be the presenting clinical symptoms in chronic Al encephalopathy, while these symptoms are absent in the acute form. Disturbed handwriting (dysgraphia) has been noted in some cases of the chronic encephalopathy, but not the acute form of Al encephalopathy. Changes in handwriting can be evaluated retrospectively in many cases, because patients in dialysis centers tend to note some of their dialysis parameters in the dialysis chart themselves. It should be stressed that many of these symptoms like relapsing speech and language abnormalities, clouding of consciousness, grimacing, tremor and asterixis, muscle jerks and seizures are common to numerous metabolic encephalopathies unrelated to Al [22, 79, 104]

Patients with acute and chronic Al encephalopathy can demonstrate unilateral or bilateral myoclonic jerks of the extremities for several months in the chronic form and days to weeks in the acute form, followed by convulsions and coma even when the exposure is terminated. It seems as if brain pathology reaches a point of no return which results in frequent epileptic insults and inevitably leads to coma and death, either due to superimposed infections like aspiration pneumonia, or pulmonary edema due to cardiac toxicity. Recovery is exceptional in severe cases. In the past, reports showed that the early stages of chronic Al encephalopathy progressed to death, mainly because the most important source of Al, namely the dialysis fluid contamination, was not recognized as such and the exposure continued for months to years [10–17, 19–42].

Fluctuation of neurological symptoms can be seen in relation to dialysis in chronic Al encephalopathy, sometimes provoked by dialysis sessions due to additional exposure with contaminated dialysate superimposed to chronic Al intoxication [28, 29].

One of the most confusing features of Al encephalopathy is the lag phase between exposure and clinical symptoms and the subsequent rapid course of the disease. The delay can be months to years in the classical form and several weeks in the acute form. Because patients are generally symptom-free or demonstrate minor symptoms like microcytic anemia, physicians are unaware of the exposure until the development of severe symptoms. Until recently it was not known that a relatively short dialysis related exposure time to Al can be fatal.

Myocardial dysfunction may arise from Al loading, but this is not substantiated as yet. Al levels have been found markedly increased in the heart in patients with chronic dialysis encephalopathy [15]. It is presumed that sudden cardiac death may practically be due to Al cardiotoxicity [27, 106]. At pathology evaluation patients Nos. 7–9 showed signs of pulmonary edema.

**Table 5.** Clinical features of acute and chronic aluminum encephalopathy

| | Acute aluminum encephalopathy [17, 23, 30, 44, 45, 62, 80, 131, 150, 154, 158] | Classical or chronic aluminum encephalopathy [10, 15–19, 26–28, 32, 35, 39–42, 77, 127, 197, 200, 273] |
|---|---|---|
| Etiology | 1) Ingestion of Al with citrate by kidney patients<br>2) Dialysate water >600–1000 μg/L<br>3) Marked rise in plasma aluminum with desferrioxamine treatment<br>4) Hemorhhagic cystitis bladder treatment with Alum<br>5) Aluminum-containing biomaterial in otosurgery | 1) Months – years of parenteral or oral Al exposure<br>2) Dialysate water 50–1000 μg/L |
| Onset | Abrupt, sudden within weeks. Between seizures apparently symptom free. Microcytic anemia and EEG changes might precede neurotoxicity | Insidious, gradual with impressive fluctuations in severity of symptoms. Microcytic anemia and EEG changes might precede neurotoxicity |
| Clinical features | I. Neurotoxicity<br>a) Communication disturbances<br>None or minimally | I. Neurotoxicity<br>a) Communication disturbances 87%<br>Speech disorder often the presenting symptom: slowing of speech, stuttering, stammering, misarticulation, 87% |
| | b) Cognition disturbances<br>None or minimally<br>Coma and death in a high percentage of cases | b) Cognition disturbances 66%<br>Concentration difficulty, Disorientation, Personality changes, Hallucinations, Dysgraphia. Coma and death in advanced cases |
| | c) Movement disorder 100%<br>Seizures 90%<br>Myoclonic jerks 30% | c) Movement disorder (without intervention 63–93%)<br>Motor apraxia, Myoclonic jerks. Seizures |
| | II. Musculoskeletal symptoms<br>Absent | II. Musculoskeletal symptoms<br>a) Fracturing osteomalacia and/or myopathy with progressive skeletal pain, proximal muscle weakness 67%, sometimes arthritis<br>b) Adynamic bone disease |
| | III. Other symptoms<br>Possibly increased susceptibility to infections<br>Possibly cardiotoxicity leading to sudden death | III. Other symptoms<br>a) Possibly increased susceptibility to infections<br>b) Possibly cardiotoxicity leading to sudden death |

| | | |
|---|---|---|
| Death | With or without continued exposure death within weeks in severe cases | With continued exposure death within 3 to 15 months. Early stages reversible with desferrioxamine treatment |
| Laboratory findings | In epidemic cases when related to water contaminated dialysate: aluminum in water used for dialysis >600–1000 µg/L | In epidemic cases when related to water contaminated dialysate: aluminum in water used for dialysis >50 µg/L |
| | Plasma Al usually >500–1000 µg/L (epidemic cases) | Plasma Al usually 100–350 µg/L |
| | Microcytic anemia (almost always) | Microcytic anemia (frequent) |
| | Hypercalciemia (depends on concomitant use of vit D, calcium preparations and calcium level in the dialysate ) | Hypercalciemia (depends on concomitant use of vit D, calcium preparations and calcium level in the dialysate) |
| | EEG changes probably in all patients (multifocal burst of slow [delta] and spike activity) | EEG changes preceding any clinical signs 4–6 months (multifocal burst of slow [delta] and spike activity) 100% |
| | Cerebrospinal fluid aluminum level relatively low (<10 µg/L) | Cerebrospinal fluid aluminum level unknown |
| | Bone aluminum level ~85 times higher than controls | Bone aluminum level ~85 times higher than controls |
| | Non-specific histological changes | Non-specific histological changes brain |
| | Brain aluminum level 0.93–1.81 µg/g dry weight (>20 times higher than normal) | Brain aluminum level 0.4–3.7 µg/g dry weight (10 times higher than normal) |
| Treatment | Seizures and myoclonic jerks do not or hardly respond on benzodiazepines | Benzodiazepines can be very effective in treating myoclonic jerks and seizures in the early stages |
| | Severe cases probably do not respond to desferrioxamine. The less symptoms, the better the outcome with earlydesferrioxamine treatment | In early stage reversible with desferrioxamine |
| | Aluminum gastroenteropathy | |
| | Acute aluminum gastroenteropathy<br>1) Oral exposure# [20, 82, 88, 89]: Unpleasant taste, mouth ulceration, joint pain, gastrointestinal disturbance, skin rashes, fatigue, nausea and vomiting, itching and sore eyes<br>2) Peritoneal exposure [81, 83]: nausea, vomiting, cramp like abdominal pain and general malaise within days | Chronic aluminum gastroenteropathy probably nonexistent |

# Other agents involved, could have contributed to, or could have been responsible for to the symptoms.
None or minimally: Either severe encephalopathy or relatively symptom free.

Al is also present in the synovial fluid and arthritis-like symptoms have been reported in patients with Al bone disease, possibly due to an inflammatory effect on the synovial cells [107, 108]. We did not see signs of arthritis shortly after the intoxication, but after several months of treatment with desferrioxamine, three patients developed a possible transient synovitis with hydrops in the knee joints.

Although Al bone disease can occur in any patient intoxicated by Al, the risk is greater in diabetics. This may be related to a lower than normal bone formation rate, an abnormality that has be demonstrated in type 1 diabetics prior to the onset of clinical renal disease [109, 110]. The risk for developing acute Al encephalopathy also seems to be increased in diabetic patients; this might be related to a decreased storage capacity of the bones [17].

## 3.4 Diagnostic Parameters

### 3.4.1 *S-Al Levels*

It should be stressed that testing of Al in body fluids can be easily disturbed by contamination, due to the ubiquity of Al. Especially talcum in medical gloves, and glass tubes are known to be responsible for falsely elevated values. Protocols have been developed to avoid Al contamination and Al-free polypropylene and polystyrene tubes prerinsed with 1 N $HNO_3$ are used for this purpose [111]. In patients with Al exposure through Al containing biomaterial in otosurgery, s-Al levels can be below 300 μg/L [112–116]. In all other intoxications resulting in acute Al encephalopathy s-Al levels are invariably well above 300–500 μg/L.

### 3.4.2 *Other Serum Determinants*

Especially in Al bone disease hypercalcemia, suppressed PTH levels, and normal or slightly elevated alkaline phosphates have been reported. Hypercalcemia may develop in some hemodialysis patients with acute Al encephalopathy when they are also exposed to oral calcium and vitamin D preparations in combination with the available calcium from the dialysate [17]. It is not known whether PTH levels and serum alkaline levels change during acute Al encephalopathy.

### 3.4.3 *Liquor*

Al levels in the cerebrospinal fluid are not always helpful for the diagnosis because even at very low levels (below 10 μg/L) neurotoxic effects can be present [23, 80, 99].

### 3.4.4
### *Microcytic Anemia*

Among dialysis patients, a microcytic anemia most commonly arises due to iron deficiency but Al toxicity should be suspected in all cases where iron deficiency is unlikely. Probably most patients with acute and chronic Al intoxication develop microcytic anemia [93, 95–98, 105].

### 3.4.5
### *EEG*

Almost all patients with acute or chronic Al encephalopathy showed EEG alterations. These were characterized by multifocal bursts of slow (delta) and spike activity. Background rhythms between bursts were relatively normal in patients with chronic Al encephalopathy, in contrast to uremia and other metabolic encephalopathies where there is general slowing of the rhythmic cycle [33, 99–102, 105, 118].

# 4
# Sources of Al Exposure Leading to Acute and Chronic Al Intoxication

## 4.1
## Contamination of the Dialysate

### 4.1.1
### *Al Contamination at Drinking Water Plants*

#### *4.1.1.1*
#### *Al Use in the Drinking Water Industry*

Since at least the Roman times, alums and the related $Al_2(SO_4)_3$ have been added to drinking water to improve the appearance and it is still the most widely used coagulant for turbid water to precipitate organic matter, because the process is cheap and practical in its use. The coagulants commonly used by water system plants are alum, poly-Al chloride, ferric salt and polymer [119, 120]. In the process residual Al levels are generally low. The Environmental Protection Agency encourages utilities to meet a level of 50 µg/L Al/L where possible, but believes that "varying water quality and treatment situations" necessitate a flexible approach [121]. Therefore, and also because no health-based guideline value for Al was recommended by the World Health Organization (WHO) [122–124], large variations in Al concentrations in drinking water can be found due to variations in source water and water Al levels as high as 1600–2700 µg/L have been reported [119]. Features of Al toxicity can affect a high percentage of dialysis patients using such water for dialysate preparation when there are inadequate water purification procedures [10, 16, 17, 19, 23–26, 37, 38, 46], or when water purification procedures insufficiently protected the patients [28, 30, 39, 44]. Table 6 gives a review of the literature on dialysate contaminations with

**Table 6.** Dialysis water contaminations in dialysis centers. Comparison of serum and dialysate aluminum levels in patients without reported symptoms[a] and patients with classical[b] and acute[c] aluminum encephalopathy

| Location | Reported Al level in dialysis water (μg/L)* | Serum Al level (μg/L) | Time between intoxication and death or encephalopathy | Death | Chronic Al enc. | n= |
|---|---|---|---|---|---|---|
| Belgium [44] | 62 | 154 ± 70[a] | – | None | None | NR |
| Egypt [44] | 85 | 119 ± 99[a] | – | None | None | NR |
| Paraguay [44] | 98 | 83 ± 43[a] | – | None | None | NR |
| Denver, USA [16] | 150–350 | – | 4–6 years | – | – | – |
| Cork, Ireland [34] | 70–450 | 147–423[b] | 22–93 months | 5 | 14 | NR |
| Newcastle, UK [35] | 300–1200 | NR[b] | <2 years | 10 | NR | NR |
| Manchester, UK [12, 127] | 6–109 | 84[a] | – | None | NR | NR |
| [37, 38] | 200–1000 | NR[b] | 3–40 months | NR | 8 | 34 |
| United Kingdom [13, 14] | ≤702 | ≤756[b] | NR | None | NR | NR |
| | ≤80 | ≤300[b] | – | None | NR | NR |
| | 20–60 | ≤165[b] | – | None | NR | NR |
| Minniapolis, UK [19] | 160–1630 | NR[b] | NR | 20/22 | – | NR |
| Trent region, UK [36, 198] | ≤480 | NR[b] | 4 years | NR | 11 | 202 |
| Scotland [26, 27] | 400 ± 33.3 | NR[b] | NR | NR | None | NR |
| | ≥1000 | NR[b] | 3 months on home dialysis | 1 | NR | NR |

| | | | | | | |
|---|---|---|---|---|---|---|
| Eindhoven, The Netherlands [28, 29] | 800–1000 | NR[b] | Years | NR | 8 | NR |
| Portugal [23, 30, 44, 45, 80] | 2200 | 104–1257[c] | Weeks | 25 | None | 71 |
| USA [39] | Up to 262 (in dialysate 81–8400)* | n = 64[b] 100–642[b] | 2 years | 1 | 8 | 64 |
| Curaçao [17, table 1] | 690 | 118–1700[c] | Weeks | 9 or 10 | None | 27 |

* calculated retrospectively.
n = total exposed patients.
From the available literature it is impossible to have a proper evaluation of the aluminum levels in the water used for dialysis, the serum levels in patients and their outcome, due to insufficient data and retrospective studies. Water aluminum values can vary substantially over a short period, and some water companies may either be reluctant to release their data, or did not test aluminum on a frequent basis. There have been only two outbreaks of acute encephalopathy: one in Portugal, and one on Curaçao with similar symptomatology and a mortality above 30%. In both studies aluminum levels in the water used for dialysis was measured after the outbreak and in both studies levels above 650 μg/L were found. In several cases levels above 1000 μg/L, and retrospectively calculated up to 8400 μg/L were reported, without similar clinical symptoms. Because acute aluminum encephalopathy has been extremely rare and with regard to the fact that patients seem to tolerate extremely high serum levels (above 1000 μg/L) before acute aluminum encephalopathy develops, it seems likely that in outbreaks in hemodialysis, levels of the water used for dialysis need to be at least above 1000 to 1500 μg/L before acute aluminum encephalopathy can develop.
NR not reported.

Al, the reported serum levels when available and the outcome of the patients. These data clearly demonstrate that there is an exceptional variation of the severity of symptoms, which hardly seem to correlate with neurotoxicity or serum levels. This is in part due to the fact that reported Al levels in municipal drinking water were evaluated retrospectively and that these levels can vary substantially from month to month.

#### 4.1.1.2
#### Relatively Low Al Levels (50–75 μg/L) in Water Used for Preparation of Dialysate

A large discrepancy between dialysate and s-Al levels are found at the lower end of the spectrum, as there is no linear relationship between Al concentrations in water used to prepare the dialysate and s-Al concentrations. Al is transferred against the concentration gradient, because it binds a non-dialyzable plasma constituent, mainly transferrin. With Al dialysate-water levels of 50–75 μg/L the plasma Al level can rise by 50% to 70% during a single hemodialysis procedure. The body load of Al may thus progressively be increased in a rate of about at least 20.5 μg/min, or some 6 mg/dialysis [32]. The transport in the opposite direction can presumably occur when the plasma protein binding sites are saturated with Al, but the amounts transferred are rather small [42]. Such observations have urged dialysis facilities to keep the dialysate Al level below 5 to 10 μg/L.

#### 4.1.1.3
#### High Levels of Al (>75–200 μg/L) in Water Used for Preparation of Dialysate

Davidson et al. found that the cumulative risk of death due to dialysis dementia in patients whose water supply had a mean Al concentration >200 μg/L was significantly greater (27.9% deaths in the first 40 months) than the risk in patients whose mean water Al content was <200 μg/L (2.1% deaths in the first 490 months). The relation between the mean Al concentration in the dialysate and time to death was given in the formula [24]: The time in months from the first symptoms of dialysis dementia to death = 65−0.081 × [mean Al concentration (μg/L) in dialysis water].

It may be postulated that the severity of exposure to Al determines the pattern of morbidity. Normally, Al in plasma is bound and transported by transferrin. However, when plasma Al levels exceed 500 to 600 μg/L, transferrrin's capacity for binding Al is exceeded, and Al probably exists in plasma in combination with citrate or as a hydroxide in association with phosphate. Thus, acute severe exposure may be more likely to produce the neurological disorder and increased sensitivity to flicker, whereas chronic moderate exposure may manifest primarily as bone disease and the other less dramatic symptoms [27, 125–128]. In any case, considering the fact that there have been only two reports of outbreaks of acute Al encephalopathy in dialysis centers [17, 43–45], and s-Al levels of 1200 μg/L [44] to 1700 μg/L (Table 1) have been reported (Table 6), acute Al encephalopathy due to water contamination probably only develops when the Al level in the tap water is well above 1000 μg/L.

There are several possible reasons for the fact that some patients remain relatively symptom-free while others develop serious neurological symptoms in cases with apparent similar exposure.

a) With concomitant use of Al-containing drugs a parenteral exposure is superimposed on chronic exposure.
b) The hours of dialysis treatment with subsequent Al loading can vary substantially, as well as the "effectivity" of dialysis due to flow differences, artificial kidney characteristics, etc.
c) Large individual differences may protect patients from neurotoxicity. Large body stores for Al (bone, spleen, liver) and volumes of distribution differ among patients. Reduced storage capacity in liver failure or osteoporosis might account for large differences. High PTH levels [129–132], high iron stores [79] and high silicon [58, 59] levels may also protect patients partly due to influences on the bodily distribution of Al. Some reports have demonstrated large differences in silicon levels between patients and silicon levels can differ 100-fold in both cerebrospinal fluid and serum compared to subjects with normal renal function [20, 23, 58]. The protective effect of silicon against Al toxicity may be either by affecting the bioavailibility of Al [20, 23, 59] or reversing the Al-induced conformational changes of neurofibrillary tangles [51, 55, 60, 61]. A residual diuresis protects against Al toxicity, as it will be excreted in these patients [67] (Fig. 2).

#### 4.1.1.4
#### *Al Contamination During Drinking Water Distribution*

Only one case reported an increase of Al during drinking water distribution (also discussed in this paper), because Al leached from the cement lining of a water distribution pipe [17, 18].

#### 4.1.1.5
#### *Al Contamination by Hemodialysis Equipment*

a) In one case, fatal Al toxicity occurred in a dialysis unit in Eindhoven, the Netherlands, due to the release of Al into water from an electrical anode used to protect a boiler from corrosion [28].
b) Several reports have related the use of Redy cartridges in hemodialysis with Al exposure. Al oxide is present in the Redy cartridge and due to the increased solubility of Al in alkaline solution, Al can be released from the Redy cartridge and cross a dialyzer membrane [76, 132–135]. The risk for Al intoxication is even higher when citrate anticoagulation is used in combination with Redy dialysis. Infusion of citrate into the arterial line, followed by infusion of calcium into the venous line, has been used for regional anticoagulation during hemodialysis of patients with bleeding diatheses or heparin-induced antiplatelet antibodies. These patients are often hospitalized in intensive care units where the use of a sorbent regeneration dialysis system is convenient [134, 135].

c) An outbreak of Al intoxication in hemodialysis patients due to an Al pump was reported in 1991. In a dialysis center in Pennsylvania bicarbonate-based dialysate passed through one of two electric pumps. The electric pumps used to deliver the acid concentrate used in bicarbonate dialysis had an Al casing, casing cover, and impeller. Elevated levels of Al were found in the acid concentrate after passing through a pump [39].
d) Eight patients undergoing acetate-free biofiltration suffered aluminum intoxication. The source of this outbreak was parenteral exposition to high concentrations of aluminum in sodium bicarbonate solutions. At the peak of intoxication serum aluminum determination revealed an average value of 147.3 ± 21 μg/L. Aluminum levels in bicarbonate solutions were 400 μg/L [96].

These outbreaks demonstrates why it is essential to ensure that all fluid pathways, storage tanks, central delivery systems, and pumps are compatible with low pH fluids before converting from acetate to bicarbonate dialysis.

### 4.2 Desferrioxamine (DFO) Administration

Patients with a substantial Al burden can be detoxified with desferrioxamine. In some cases where high dosages were used patients can demonstrate seizures and myoclonia a few hours after administering the chelator. It is argued by Ackrill et al. that the rise in plasma Al levels cannot explain the rapid neurological deterioration since it begins within hours of administering DFO, whereas peak Al levels only occur 24–48 hours following DFO infusion [11]. It has been suggested that desferrioxamine has a direct neurotoxic effect on the brain. Another hypothesis suggests that after Al alters the permeability of the blood-brain barrier, and allows the aluminoxamine complex with a relatively low molecular weight of 583 daltons to cross the blood-brain barrier more readily than naturally circulating Al that is largely bound to serum transferrin and albumin especially because the DFO-Al complex is lipophilic [136]. A recent study however, makes it more likely that unchelated desferrioxamine passes the blood-brain barrier and chelates the Al present in the brain to convert it to aluminoxamine causing acute neurological symptoms in the already damaged brain [80].

### 4.3 Gastrointestinal Exposure

#### 4.3.1 *Oral Phosphate Binders*

Because of poor phosphate removal with dialysis, the use of Al or calcium phosphate binders has been required for lowering serum phosphate levels in uremic patients. A prospective, controlled study of pediatric patients demonstrated that a presumed "safe" dose of Al gels leads to Al accumulation and the current opinion is that there appears to be no safe dose of Al in chronic renal failure that is also large enough to control the plasma phosphate concentration. The dietary intake of phosphate is higher in North America than in many European and Asian countries, and the amount of Al hydroxide prescribed in dialysis patients had been high in these countries. This contributed to a considerable number of patients suffering from Al-related bone disease [137–143]. Close correlations have been observed between s-Al levels and the amount of Al-containing gels ingested in both children and adults undergoing dialysis. In children with renal failure, the ingestion of doses of Al hydroxide containing more Al than 75 mg/kg per day was associated with plasma levels of Al above 100 μg/L, and features of Al bone toxicity were common [78, 137, 143–145]. In uremic patients life-threatening Al encephalopathy has been observed when Al hydroxide has been administered with sodium citrate [146] or calcium citrate [28]. Non-dialyzed uremic patients showing Al encephalopathy had received the combination of oral Al with citrate compounds [146]. Citrate is known to enhance the passage of Al over the intestinal wall and the blood-brain barrier and to contribute to the development of toxicity [3, 147]. Uremic patients prescribed medications containing citrate (calcium citrate, Shohl's solution) or Alka-Seltzer, along with those medications containing Al (Basaljel, Amphogel, Carafate) are particularly prone to Al accumulation [148].

## 4.4
## Bladder Irrigation with Al Sulfate

Al toxicity also has been reported from intravesical bladder irrigation with Al sulfate (alum). Alum is an astringent and controls bladder hemorrhage. In the past it was regarded to have a non-significant systemic absorption, but probably a breach of endothelial integrity secondary to hemorrhagic cystitis allows Al to leach across the vesical mucosa, contributing to significantly tissue loading. As the intact renal function is essential for a rapid disposal of a parenteral Al load, most reported instances of encephalopathy after alum irrigation have occurred in patients with compromised renal function. Manifestations of encephalopathy appeared after continuous irrigation with 1% alum for at least 2 days and are associated with elevated s-Al concentrations. Patients are treated with alum with up to 30 liters of irrigant over a 24 h period [149]. Probably most cases involve acute Al encephalopathy with serum levels of 135 to 436 μg/L and a high mortality [149–155]. The symptoms are similar to that of acute Al encephalopathy in dialysis patients with convulsions and coma but interestingly, the lag phase of 2 to 8 days [150], is much shorter than reported in hemodialysis outbreaks of acute Al encephalopathy. This is possibly related to both the rapidity and magnitude of Al loading, causing large body stores for Al (bone, spleen, liver) to be partially bypassed.

Some authors claim to have had patients with Al neurotoxicity symptoms when the highest s-Al concentration was 17 to 22 μg/L [156, 157]. It seems very unlikely that these patients indeed suffered from Al encephalopathy because the patients did not show the typical symptoms and s-Al levels below 100 µg/L are generally considered unlikely to cause encephalopathy in hemodialysis patients [20]. In this case several other complicated metabolic abnormalities, in retrospect, were probably responsible for the neurological symptoms [150]. Nevertheless, alum should not be employed as a bladder irrigant, especially not in patients with acute or chronic renal failure.

## 4.5 Al Containing Biomaterial (Cement) in Otosurgery

Ionomeric cements are used routinely in otosurgery, especially in maxillofacial or skull bone-cementoplasty by an Al-containing biomaterial and canal reconstructions where the proximity to the facial nerve is evident. Cement, however, is not inert, and Al is an important constituent of cement. The literature contains reports of diffusion of Al ions, which can reach toxic levels in tissue fluid and adjacent bone as the cement hardens. During setting and hardening, the hybrid bone substitute ionomeric cement achieves a stable and durable bond with the apatite of the adjacent bone without interpository soft tissue. A close contact between an Al-containing cement and the cerebrospinal fluid results in the release of Al ions which may reach critical levels as high as 3000 µg/L [23, 128]. This has resulted in a fatal (sub)acute Al encephalopathy in patients without renal failure. After a latent phase of up to six weeks, patients suffer from loss of consciousness, myoclonic jerks, and persistent grand mal seizures [158–160]. A direct toxic effect on peripheral nerve transmission has been reported when ionomeric cements were introduced into near-nerve anatomic locations, mainly the facial nerve, following surgical reconstruction of the posterior ear canal with ionomeric cement. Paralysis developed in some cases postoperatively within two to three weeks [115]. Except in cases where individual peripheral nerves adjacent to the biomaterial showed signs of dysfunction, symptoms of post-otosurgery Al encephalopathy are quite similar to acute Al encephalopathy with other exposure routes. They include convulsions and myoclonus of the extremities but also stupor, and mutism. In some patients there was no clinical improvement after the biomaterial was removed. Because of these cases the use of one of these cement compounds (Ionocem) was stopped in France by order of the Direction Généerale de la Santé in 1994. It was suggested that all patients operated on with these products should have serum and cerebral fluid Al analysis and electroencephalograms and the biomaterial should be removed when abnormal values were noted [112].

## 4.6 Peritoneal Dialysis and Acute Al Gastroenteropathy

Acute Al gastroenteropathy with symptoms of nausea, vomiting, cramp-like abdominal pain and general malaise can follow after accidental Al contamination of the peritoneal dialysis fluid [81, 161]. It seems as if Al has a direct toxic effect on the outer epithelium of the gastrointestinal tract.

## 4.7 Other Sources of Al Exposure not Related to Acute Al Intoxications

### 4.7.1 *Parenteral Nutrition and Oral Milk Based Formula in Children*

Pediatric patients are particularly at risk for Al neurotoxicity. The development of the brain occurs in the first year of life. High permeability of the immature blood-brain barrier to Al, the increased uptake of Al via a relatively poorly developed gastrointestinal tract, immature formation and function of the kidneys and high Al to body ratios probably all contribute to the Al toxicity [144, 145, 162–165]. Children not on dialysis were intoxicated by Al-containing phosphate binders and developed encephalopathy [78, 162, 164, 166, 167].

A study of newborn and premature infants found that the intake from milk and formulas for preterm infants is about 0.03 mg Al/day [168, 169]. Several cases were reported to develop neurological symptoms with elevated Al tissue levels receiving no dialysis and no Al-containing phosphate binders, but Al-containing milk based formula [165, 170]. The Al content of breast milk is very small (about 20 μg/L), and formulas based on cow's milk contain about ten times as much Al. The highest amount of Al have been reported from soya-based formulas for lactose intolerance (200 μg/L to 1000 μg/L), but its toxicity risk is unknown [171, 172].

In preterm infants, prolonged intravenous feeding with solutions containing Al is associated with impaired neurological development. These studies have methodological problems that may influence the results, but nevertheless acute Al encephalopathy with seizures and myoclonia has not been reported. Low dose of Al, present in parenteral nutrition formula, can produce marked portal inflammation correlating with the duration of exposure and the amount of Al accumulated in the liver [173]. As a result of the concern about Al in total parenteral nutrition solutions, the United States Food and Drug Administration now requires that Al concentrations be under 25 μg/L in large volume parenteral drug products [174].

### 4.7.2 *Drinking Water*

In recent study the relative bioavailibility of Al naturally present in food and in alum-treated drinking water (ATW) were determined using 29 healthy volunteers, who drank, during two-day periods, ATW or pure water, with and without citrate, while on a controlled diet. Only 1–2 percent of the daily intake of Al came from ATW and only 0.3 to 0.4 percent of the Al in ATW was absorbed by the body, which was the same percentage as that absorbed from food. It was estimated that drinking 1.6 L/d of ATW containing 140 μg/L Al would contribute only 0.4–1.1 percent of the lifetime body burden of Al [2]. However, although Al is very poorly absorbed in the gastrointestinal tract, and the possibility that some Al fractions present in drinking water may be particularly bioavailable cannot be dismissed at present, especially when exposure starts from childhood [61]. In addition, oral administration of $^{26}$Al to rats resulted in higher brain $^{26}$Al concentrations [176]. Recent reports on the risk of Al levels below 100–200 μg/L in drinking water do not indicate an important risk for the normal population [122–124, 175].

### 4.7.3
### *Inhalatory Exposure*

Most of the Al absorbed from the respiratory tract accumulates in the lungs. Pulmonary lesions have been described in employees of Al processing or manufacturing industries and encephalopathy after Al inhalation has been reported. Al is widely distributed and has many industrial uses, and toxicity from occupational exposure is assumed to be extremely rare [2, 177]. Nevertheless, a recent study investigating adverse effects on the central nervous system of Al welders found an Al-exposure-related increase in blood and urine Al concentrations, deficits in neuropsychological test performance and mild diffuse EEG abnormalities. Therefore, the potential for Al-induced neurotoxicity in those occupationally exposed to Al fumes may be greater than previously suspected [177].

*Aerosol antiperspirants* can be absorbed systematically via the lungs and airways and through the nasal-olfactory route. There is as yet insufficient information available to determine the possibility of Al accumulation in the brain [178].

### 4.7.4
### *Dermal Exposure*

*Blepharopigmentation* with Al-silicate has been introduced during the last decade as a technique for creating a permanent line along the eyelid margin, thus simulating a cosmetic eyeliner. Patients can have a delayed hypersensitivity granulomatous reaction [179].

*Dermal persperants*: in a preliminary study the dermal absorption of Al from antiperspirants have been estimated to be 0.03–4 μg/L, or 0.012% in one study. The clinical significance is unknown [180].

*Vaccins adjuvants*: the purpose of a vaccine adjuvant is to enhance the immune response of the vaccine. An adjuvanted vaccine will tend to have a higher, earlier, and longer-lasting immune response than a non-adjuvanted vaccine. Until recently, the only vaccine adjuvants approved for human use were Al salts generally referred to as alum. Commonly used Al-adjuvanted vaccines include: diphtheria, tetanus, hepatitis, lyme, rabies and anthrax [181]. Known side effects from Al adjuvants include short-term effects such as swelling, inflammation and granulomas [182] but recently there has been a controversial finding that macrophagic myofasciitis is associated with Al-containing adjuvants [168, 182]. Although the dissolution of the Al adjuvants can take 4 to 18 months, according to animal studies, probably very little Al is incorporated into the brain [168].

### 4.7.5
### *Intranasal Absorption*

There are no data that could be used to estimate Al bioavailibility following intranasal exposure to indicate whether this represents a significant route of exposure [4], but toxicity through this route has not been reported.

### 4.7.6
### *Low, or Absent Toxicity of Other Sources of Al Exposure in Humans in Normal Circumstances*

Commercially, Al salts are added to foods, such as frozen strawberries, maraschino cherries, and pickles, to improve their appearance. $Al^{3+}$ salts are often added to processed grain products, salt, cheeses and beer [183]. Plants can accumulate $Al^{3+}$ especially in acidic soils, and evidently detoxify the $Al^{3+}$ by possessing basic fluids or organic chelators. Especially tea is known to store $Al^{3+}$ in cell vacuoles in older leaves separate from metabolically active parts of the plant. As much as 3% of $Al^{3+}$ can be found in older tea leaves, while only 0.01% in younger ones, a 300-fold difference. Commercial teas yield up to 100 ppm $Al^{3+}$ in a drink, but much higher Al concentrations can be found in certain species of leaves [184, 185]. Adding milk to form the insoluble phosphate can reduce gastrointestinal absorption of Al from tea but by addition of lemon substantial amounts of an Al citrate complex will be absorbed. Numerous other food substances were described as a possible – most likely harmless – source of Al, like small amounts of dissolved Al from cookware, fruit juices, coffee and baking powder [186, 187]. Some foods contain large amounts of Al as a result of processing. Levels of up to 4 mg have been reported to be present in one serving of food [186]. Although the major source of Al in drugs remains the use of Al hydroxide phosphate binders in renal patients, some common oral pharmaceutical products can contain high amounts of Al, such as calcium supplements (12 mg/day), buffered aspirin (10–20 mg/tablet), and Al-containing antacids (50 mg/tablet) [140, 188]. Much effort is being taken to reduce the Al content of the above products or to switch to alternative products which contain less Al [20].

# 5 Pathogenetic Mechanisms of Aluminum Intoxications

## 5.1 Speciation, Biokinetics and Metabolism

### 5.1.1 *Al Speciation*

Despite several decades of clinical experience and animal study, the pathophysiology of brain encephalopathy is far from understood. One of the main reasons for the complexity of the research on Al is the difficult interaction with molecules found in biological systems. In most natural systems, a small fraction of Al is found as the simple $Al^{3+}$ aquo ion. Thus Al absorption, excretion, tissue retention, and deposition will all depend on the properties of the $Al^{3+}$ complexes formed with biological ligands. Unfortunately, the search for an accurate description of Al complexation equilibria and kinetics has been consistently frustrated by the tendency of both free $Al^{3+}$ ion and simple Al complexes to hydrolyze at or below neutral pH.

An important issue in the toxicity of Al is the solubility and speciation of inorganic Al, which are strongly pH dependent, whereas the solubility of organically complexed Al is correlated with the level of dissolved organic matter. In aqueous circumstances with low concentrations of dissolved organic compounds, the dependence of dissolved Al on pH resembles, on a logarithmic scale, a parabola with a sharp solubility minimum at around pH 6.5. In water more acid than pH 5, $Al^{3+}$ exists as an octahedral hexahydrate, $Al(H_2O)_6^{3+}$, usually abbreviated as $Al^{3+}$ and referred to as free Al. As a solution becomes less acidic, $Al(H_2O)_6^{3+}$ undergoes successive deprotonations to yield $Al(OH)^{2+}$, $Al(OH)_2^+$ and soluble $Al(OH)_3$, with a decreasing and variable number of water molecules. At pH > 6.2 tetrahedral aluminate, $Al(OH)_4^-$, is the primary soluble $Al^{3+}$ species. The Al levels in the water will be related to the pH conditions with Al hydroxide solubility equilibria but in addition to the hydroxide ion, the Al ion will also form complexes with inorganic ligands such as fluoride ($F^-$), sulfate ($SO_4^{2-}$) and silicate ($HSiO_4^{3-}$). Therefore, the actual chemical species of Al is strongly predisposing for the availability of Al for intestinal uptake as well as its metabolic behavior and toxicity [189–193]. The mixed Al(ligand) – hydroxide complexes are the rule, rather than the exception – and hydroxo-bridged polynuclear complexes are common [193]. Metal hydrophilic species show different neuritogenic properties indicating the ability of $Al^{3+}$, when diversely coordinated, to produce different biological effects [129]. Two main $Al^{3+}$ species are recognized in aqueous solutions: the hexahydrate $Al^{3+}$ at pH < 5.5 and the tetrahedral aluminate at pH > 6.2. In blood plasma, citrate is the main small molecule carrier and transferrin the main protein carrier of $Al^{3+}$. In fluids where the concentrations of these two ligands are low, nucleoside di- and triphosphates become $Al^{3+}$ binders. Under these conditions $Al^{3+}$ easily displaces $Mg^{2+}$ from nucleotides. When all three classes of ligands are at low concentrations, catecholamines become likely $Al^{3+}$ binders. In biological systems $Al^{3+}$ will associate with oxygen donor ligands, at least some of which should be anionic to counter the 3+ charge of the cation. Carboxylate and phosphate groups, nucleotides, and polynucleotides meet this description. Double-helical DNA binds $Al^{3+}$ weakly. $Al^{3+}$ forms weak complexes with amines and sulfhydryl ligands. With multidentate amino-carboxylate ligands such as EDTA, $Al^{3+}$ forms strong complexes. $Al^{3+}$ in the cell nucleus probably binds to nucleotides or phosphorylated proteins [190].

## 5.1.2
## *Biokinetics*

### *5.1.2.1*
### *Absorption and Excretion*

In man, without excessive exposure to Al and a normal renal function, the total body burden of Al is 30 to 35 mg, and the intake of Al in foods is 2 to 3 mg/day, with only a small percentage absorbed [15, 129]. The lungs contain the largest amount of Al because dust is trapped in pulmonary macrophages and its concentration increases slowly with age. After the lungs the highest concentration of Al is found in the liver, bone, spleen, myocardium and parathyroid cells and their concentration is 1 to 4 mg/kg dry weight. In these subjects bone and muscle account for about 40% of total body Al and the lungs for 12% [129, 194]. Because trace metals are indestructible, chronic accumulation of the element will occur in patients with end-stage renal failure, because the major elimination route, i.e., the kidney is absent. This results in a much higher Al burden, ranging between 220 and 1600 mg. In these patients also a different distribution is observed with bone and liver as the major storage sites with comparable Al concentrations [129]. Most ingested Al is eliminated in the stool within several days and the Al bioavailibility from food is ~0.1%. The remaining Al, which enters the blood via the intestinal system, is then eliminated in urine. Daily urinary Al excretion has been estimated to be 4–12 μg [4]. About 85% of the intravenously injected Al and gallium citrates is normally removed in two weeks. The amount of Al that is not eliminated in the urine is then retained in the body by deposition in tissue and removed very slowly. An estimated terminal $t_{1/2}$ in one human being who received intravenous $^{26}Al$ was 7 years [194, 195].

For many years, there has been some controversy about the quantitation of intestinal absorption of Al. Because of the lack of a suitable isotope and a sensitive technique of analysis, in the past Al have been studied indirectly using analogs such as $^{67}Ga$ ($t_{1/2}$ = 78 h). Recently, with the development of accelerator mass spectrometry (AMS), it has become possible to use the artificially produced radionuclide of Al, Al 26 ($^{26}Al$), ($t_{1/2} = 7.16 \times 10(5)$ years) [84]. With this technique, literature data on fractional gastrointestinal absorption still varied between 0.04 and 1% [59, 87]. The fractional intestinal absorption of trace oral doses of Al hydroxide is at least 0.1% (compared with the previous estimate of 0.01% using large $^{27}Al$ oral loads). Absorption of Al citrate given alone is significantly greater (0.7%) and is further increased to 5%

by the accompanying sodium citrate, consistent with an enhancing effect of added citrate upon mucosal Al permeability. Al maltolate absorption approximates that of Al hydroxide (0.1%) [129, 194]. Although no single explanation of the exact mechanism of gastrointestinal absorption has emerged, it is believed that intestinal absorption of Al includes both paracellular pathways along enterocytes and through tight junctions by passive processes as well as transcellular pathways through the enterocytes involving both active and passive processes [1]. In normal subjects there is little or no increase of tissue Al content (except in the lungs) with aging; this suggest that the body is in neutral balance. The factors that affect the absorption of Al include uremia, dietary factors such as citrate, a low gastric pH, fasting, and also a lesser effect on absorption has been reported from lactate, parathyroid hormone and vitamin D. When Al enters the body, the kidney is the major organ for excretion and in renal failure there will be a widespread deposition in extracerebral organ systems of patients with Al overload and the bone and liver serve as the major storage organs [194]. Sequestration of Al in bone and liver protects the brain from Al toxicity [195, 196]. This is supported by the fact that conditions associated with a negative calcium balance, such as immobilization, surgery, corticosteroid therapy, and hypophosphatemia, may precipitate the onset of the disease [197]. In the presence of high PTH levels there will be an increased bone uptake of Al [198, 199]. On the other hand, Al also influences the parathyroid negatively and lowers the parat hormone secretion, which reduces bone turnover [200]. An open compartmental model for describing Al biokinetics after p.o. administration of $^{26}Al$ has been presented in one study, with a central compartment consisting of transferrin- and citrate-bound Al in blood plasma and interstitial fluid, and three peripheral compartments for organs, muscles and bones and the gastrointestinal tract. The rate constants describing the transport of Al are normalized to an estimated plasma volume and do not depend on the size of the individual [201]. Other routes of absorption (inhalatory, dermal and intranasal) are discussed above.

### *5.1.2.2*
### *Distribution*

#### 5.1.2.2.1
#### Metabolism

*Citrate*: Since it occurs to the extent of about 0.1 nM in the blood plasma, citrate becomes the pre-eminent small molecule plasma binder of a metal ion like $Al^{3+}$ that prefers oxygen donor ligands. Citrate complexion of $Al^{3+}$ in the upper gastrointestinal tract with a pH of 2 to 5, may form a zero-charged complex that may pass through membranes. Sodium citrate as well as calcium citrate may cause a 100-fold enhancement of gastrointestinal Al absorption. The adsorption of Al from $Al(OH)_3$-based antacid rises substantially upon intake of citrate, and therefore these antacids should not be taken with citrus fruit or juices [190].

*Protein binding*: Although transferrin is generally acknowledged to be the serum protein binder of Al, also other proteins such as albumin, albindin [205] and de novo synthesized proteins [72] have been suggested as potential s-Al binders. The common albumin and globulin proteins of the plasma bind metal ions such as $Al^{3+}$ only weakly and non-specifically. With a pair of sites that avidly binds $Fe^{2+}$, transferrin stands as the leading plasma protein for $Al^{3+}$ binding and carrier in the plasma [190]. The cerebrospinal fluid lacks any transferrin and contains only about 2 μM citrate. The transport of Al by transferrin favors the delivery of Al to cells with numerous transferrin receptors, such as hepatocytes and red blood cells precursors [23].

*Catecholamines*: By binding to the catechol of cathecholamines, trace amounts of $Al^{3+}$ may possibly disrupt many neurochemical processes [190].

### 5.1.2.3 Nucleotides and Organic Phosphates

Al behaves as an extremely potent genotoxic agent and specific genetic sub-compartments are selectively targeted. Numerous molecular genetic mechanisms have been described, involving Al neurotoxicity towards neuronal and glial cell nuclei and brain specific gene transcription. In spite of a considerable body of scientific evidence, the precise pathophysiology remains unclear. One emerging hypothesis is that Al is strongly attracted to the phosphates of nucleic acids (and nucleotide triphosphates) especially rich in adenine (A) and thymine (T) residues, so brain RNA and DNA sequences containing a high molar percentage of A + T nucleotides may be particularly at risk for the deleterious effects of Al. In particular, the high content of intranuclear phosphate (approaching 60 mM), the normally decondensed or extensive euchromatization characteristic of neocortical neuronal nuclei, their intrinsically high rate of transcriptional output and their generation of complex RNA message populations required for normal brain cell function may specifically target neutral cells for many of Al's genotoxic effects [66]. In experimental Al neurotoxicity studies, especially when acute high dose of Al were applied, Al targets numerous cytoplasmic and nuclear structures. For example, a single stereotactic injection of 100 μg/L of a 1 percent Al lactate solution into rabbit cerebral ventricles results, within 3 hours, in decreased levels of RNA messages encoding the key cytoskeletal components alpha-tubulin, beta-actin and the light chain of the neuron-specific neurofilament protein [66, 204]. This observation would be in accordance with, though not explain, the development of neurofibrillary pathology as observed in experimental Al intoxications and in Alzheimer's disease.

Another interesting speculation is that Al is able to exchange other key cations like iron, magnesium and zinc from binding sites necessary for normal molecular genetic function of transcription factor proteins and thereby play an ancillary role for further disrupting gene transcription mechanisms [66]. After calcium and iron, zinc is the third most abundant trace element in the mammalian brain, and has been receiving a lot of attention lately because of its potential role in the development of human neuropathology [188, 206, 207].

Excessive synaptic zinc release and postsynaptic zinc accumulation may contribute to neuronal loss associated with certain acute brain injury conditions including transient global ischemia and epilepsy. In the normal brain, zinc is thought to provide highly specialized neuromodulatory, neuroprotective and neurosecretory functions and is intimately involved with structural, regulatory and enzymatic proteins that provide these functions [188]. Bioinformatics analysis of newly sequenced human DNA from the human genome project suggests that up to 1 percent of total human coding DNA, or up to 1000 genes, encoding zinc-requiring transcription factors [207, 208] and therefore Al has the potential to interact with many important zinc-containing elements of the brain's gene expression system [66].

## 5.2
## Neuropathological Mechanisms

The published descriptions of the neuropathologic changes of chronic Al encephalopathy in the past have been remarkably brief, absent or non-specific, and no data at all are available from the neuropathologic changes in acute Al encephalopathy in humans. In the brain significant Al content was observed in the hippocampus, the occipito-parietal cortex, the cerebellum, and the striatum. This topographic distribution correlated with the clinical defects in higher cortical functions, including aphasia. Spongy changes consisting of vacuoles in the neuropil and inside nerve cell bodies were found and the vacuoles were located mainly in dendrites and astrocytic processes. Thus, Al encephalopathy is a disease of neurons and astrocytes of the cerebral cortex [103, 209–211]. None of these abnormalities found post-mortem were sufficient to account for the patients' neurological conditions [25, 211]. In animal studies most pronounced changes in the brain were observed in the hippocampus [212].

The development of a new silver-staining technique by Reusche allowed the visualization of characteristic patterns and distinct morphological changes of chronic Al encephalopathy in humans. Central nervous tissue (CNS) and peripheral organs of 50 autopsy cases with chronic renal failure (CRF) and dialysis treatment were evaluated for Al-containing argyrophilic inclusions using the Howell and Black method modified by Reusche [51, 195, 213]. Al encephalopathy morphology is characterized by lysosome-derived intracytoplasmic, Al-containing, pathognomonic, argyrophilic inclusions in choroid plexus epithelia, cortical glia and neurons. The most sensitive structures for CNS deposits are choroid epithelia, followed by glial cells and neurons. Autonomic ganglia, heart, ovary/testis, parathyroid, adrenal, and pituitary demonstrated reliably peripheral deposits [213, 214]. Glial cells presented massive silver-stained deposits, which were restricted to the gray matter. Finally, neurons revealed numerous fine-granular black inclusions, scattered throughout the cytoplasm. Brainstem nuclei were primarily affected, but neurons within the cortex, subcortical gray matter and spinal cord were also involved to various degrees; inclusions were not evident in the nucleus dentatus and the oliva inferior. Vessel-related deposits were found frequently. By electron microscopy the cytoplasm of neurons was filled either with large amounts of small electron-dense granules, or with lipofuscin granules, containing numerous irregular, non-membrane-bound inclusions. Massive electron-dense depositions were seen in the cytoplasm of the choroidal epithelia and in proximity to nuclei of cortical astro- and oligodendroglia. These deposits remain for many years and even after renal transplantation, with termination of drug-related Al intake and normalized renal Al excretion, the Al-induced argyrophilic degradation products remained in the cellular cytoplasm in unchanged fashion up to 10 years [92].

Many investigators have also observed neurofibrillary degeneration changes that were regarded to be similar to neurofibrillary tangles in Alzheimer's disease and the hallmark of the experimental encephalopathy induced by Al in rabbits, cats and ferrets is neurofibrillary tangles [50, 60, 70, 215]. Al-induced tangles are due to the accumulation of 10 nm neurofilaments in the body and processes (axon and dendrites) of neurons [48–50]. Neurofilaments begin to accumulate a few hours after intracisternal injection of Al [21]. The diffuse distribution of neurofibrillary pathology suggests that nearly all nerve cell populations are prone to the toxic action of Al, the most vulnerable being neurons of the anterior horns of the spinal cord, of the reticular formation and of the basal forebrain, the Purkinje cell system and the pyramidal neurons of the cerebral cortex. These studies have supported the theory concerning Al as an etiological agent for cerebral changes in Alzheimer's disease [70]. Recently, however, skepticism has developed about the similarity of these neurofibrillary tangles in Al encephalopathy and Alzheimer's disease. In a study using a new silver staining technique, the described neuronal changes and, in particular, alterations of choroidal epithelium and glia were demonstrated to be completely different from the characteristic plaques and tangles in senile dementia of Alzheimer type [51, 92]. This assumption is also supported by the fact that neurofibrillary deposits are detected in the proximal end of the axon only one day after experimental administration of Al, while it takes several weeks before neurofibrillary degeneration is seen near the nucleus. During this latency period the neurons show progressive loss of excitability. Therefore, cell damage is conceivably not caused by the presence of fibrillary deposits, per se, but by the loss of the normal functions of neurotubules and neurofilaments [31].

## 5.3 Theories of Pathophysiology of Aluminum Encephalopathy

### 5.3.1 *Blood-Brain Barrier (BBB)*

The brain is defended by a continuous layer of cell membranes, the blood-brain barrier (BBB), which not only hinders the entry of large molecular mass substances into the central nerve system, but additionally regulates all the exchanges of nutrients, hormones, toxins and therapeutic agents that enter the CNS from the blood (to see a diagram visit http://www.sfn.org/briefings/blood-brain.html) [216–218]. When Al enters the systemic circulation it binds to transferrin, the predominant Al species in plasma, and results in a rapid entering of the brain. It has been suggested that Al can alter the BBB permeability and alter the flux of molecules and ions into and out of the brain [217], but others state that Al uptake into the brain is not dependent upon alterations to BBB permeability [218–220]. This brain Al entry could be mediated by transferrin-receptor-mediated endocytosis and is enhanced when Al binds to citrate (TfR-ME) (Fig. 3) [4, 221, 222]. Al can penetrate into the extracellular space of the brain frontal cortex and ventral hippocampus within 20 minutes [217]. Although some Al that enters the brain is rapidly refluxed, it is suggested that a fraction enters brain compartments within 24 h from which it is only very slowly eliminated [223]. The $t_{1/2}$ of $^{26}Al$ in rat brain was >100 days following intravenous $^{26}Al$ transferring dosing [222–224]. Although neurotoxicity may be the most striking form of Al-related toxicity, the brain (and also blood) in fact contains the lowest levels of Al overload [31, 38, 151, 225–227]. Transferrin receptors are expressed on blood vessels, large neurons in the cortex, striatum and hippocampus as well as oligodentrocytes and astrocytes [228] but in the cerebrospinal fluid Al will preferentially bind to citrate due to the fact that the molar concentration of citrate in cerebrospinal fluid is up to 900-fold higher than that of transferrin [23, 218].

### 5.3.2
### *Brain Al Homeostasis and Cellular Mechanisms*

Al has been found in glia (mainly astrocytes, oligodendrocytes and microglia) [229] and in the perinuclear endoplasmatic reticulum of neurons [230, 231]. Disruption of glial cell function by Al results in the accumulation of unwanted, probably cytotoxic, debris and modulation of synaptic transmission and neurone-glial signaling [232]. $Al^{3+}$ binds almost $10^7$ more strongly to $ATP^{4-}$ than does $Mg^{2+}$ [189] and forms a stable complex, which is more stable than a complex with Mg [233]. $Mg^{2+}$ is associated with phosphate groups and $Al^{3+}$ can compete with $Mg^{2+}$ for the phosphate sites. In the brain ATP acts upon extracellular inotropic (P2X) and metabotropic (P2Y) receptors to optimize the activities of neurotransmitters including glutamate, gamma-aminobutyric acid (GABA) and acetylcholine [234–236]. It was also suggested that a disturbance of neurotransmitter metabolism in the brain as a result of Al inhibition of dihydropteridine reductase is responsible for the neurotoxicity [237]. To date, 14 different P2 receptors have been identified in the brain. Some of them might be released as a complex with Mg, and in these cases Al will be intracellular in competition to form Al-ATP instead of Mg-ATP. Al-ATP might act upon muscarinic receptors to potentiate the negative feedback controlling the release of acetylcholine into the synaptic cleft, causing deficits in neurotransmitter stimulation [238]. Another mechanism, which might explain the neurotoxic and other actions of Al, is the

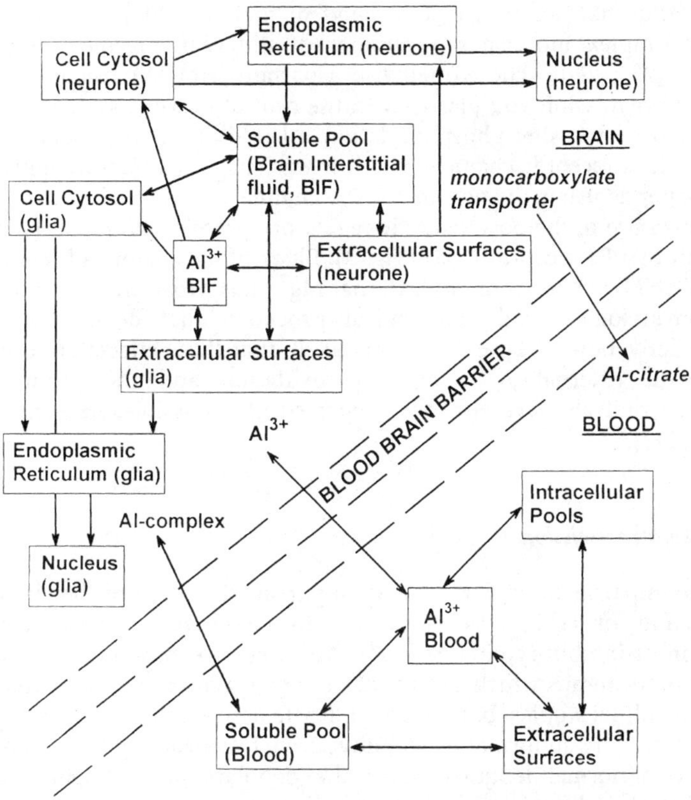


**Fig. 3.** The dynamic partitioning of aluminum in the brain results in three (extracellular, surface associated and intracellular) significant sources of biologically available aluminum. In the blood, aluminum is bound by proteins, such as transferrin, and a number of lower molecular weight ligands which could include small peptides, nucleotides, nucleic acids, citrate, phosphate and silicic acid. In BIF (Brain Interstitial Fluid), aluminum is bound by some of the lower molecular weight ligands found in blood as well as neurotransmitters such as glutamate and GABA. Aluminum is associated with the surfaces of the BBB (Blood Brain Barrier), glial cells and neurones. It may be bound by the phosphate headgroups of lipids which act as sites for the nucleation and aggregation of aluminum. Intracellular aluminum may be found bound to ATP or in endoplasmic reticulum and Golgi in close proximity to the nucleus. Aluminum may also be found in nuclear chromatin. There is continuous exchange of aluminum between intracellular and extracellular compartments and the predominant exchange pathway will change in tandem with changes in brain physiology. With permission, from: Exley C. (1999) "A molecular mechanism of aluminum-induced Alzheimer's disease?" J Inorg Biochem: 76: 133–140; Elsevier Science [228]

interaction with calmodulin. One of the most abundant and versatile $Ca^{2+}$-binding proteins, calmodulin regulates a large number of cellular processes and target proteins in response to $Ca^{2+}$ signaling. Calmodulin is found in all eukaryotic cells. It couples the intracellular $Ca^{2+}$ signal to many essential cellular events by binding and regulating activities of more than 40 different proteins and enzymes in a $Ca^{2+}$-dependent manner [207]. The calmodulin-calcium complex modulates a number of different enzymes and cellular processes [239, 240]. The *N*-methyl-D-aspartate (NMDA) receptor mediates synaptic transmission and plasticity in the central nervous system (CNS) and is regulated by tyrosine phosphorylation. Al inhibits Ca-dependent inactivation of NMDA receptor channels [241]. It was suggested that the inhibition of the Ca-dependent inactivation of NMDA channels by Al could occur through the stabilization of the post-synaptic regulatory protein that might be possibly a sub-unit of a P2 receptor [238] (Fig. 3). Block of the channel of *N*-methyl- D-aspartate (NMDA) receptors by external $Mg^{2+}$ has broad implications for the many physiological and pathological processes that depend on NMDA receptor activation [241] and it seems likely that these effects are even more severe with external $Al^{3+}$. Lipid peroxidation and the production of superoxide radicals have also been reported as a possible mechanism of Al toxicity [242].

### 5.3.3
### *Al and Neurotransmission*

Cognitive impairments due to Al may in part be the result of the alteration of the function of GABA receptors due to cognitive impairment through disruption of inhibitory circuits [243]. Al lowers the excitability of the nerve cells of hippocampus which can result in convulsions. Synaptic currents were normal in animal studies, but the possibility to activate the spike discharge was less effective. The abnormal excitability of such neurons *in vitro* is probably related to abnormal lengthening of the depolarizing after-potential, with reduced post-discharge depolarization [47], and to reduce electronic length of the cell as well [244]. Another typical feature of Al encephalopathy is the progression of the disease with an increasing number of clinical epileptogenic features until coma and death ensues. This might be related to a increasing loss of neurons. Neuropathological studies show a decreased number of tangle-bearing neurons and severe nerve cell loss, especially in anterior horns and hippocampus, and behavior studies reminiscent of temporal lobe epilepsy, may be due to the atrophy of the hippocampus [21, 22].

### 5.3.4
### *Delayed Neurotoxicity*

Cellular and molecular mechanisms of neurotoxicity are also influenced by the fact that neurons are postmitotic and do not divide. Thus, the capacity for replacement of damaged cells does not exist in the nervous system, whereas most other organ systems have a well-established capacity for regeneration. Many neurotoxins can cause encephalopathy and an important concept in neurotoxicology is the delayed manifestation of symptoms sometimes up to years after the exposure started. Several agents show a lag time between exposure and neurotoxicity. Examples are the organophosphate chemical warfare agents [245], bismuth intoxications [246] and methylmercury intoxication [247] with a lag phase of weeks to months. One of the longest delays in time between exposure and neurotoxicity however, seems to be in the case of Al encephalopathy, where it can take several years to develop. In one case report a 14-year-old boy was described who sustained a skull wound as a result of a hand-grenade explosion. This resulted in implementation of a fragment of metallic Al into the left occipital brain. Fifteen years later, at age 29, he developed seizures, mental disturbances and language difficulty. After a gradual deterioration he died at age 34 in status epilepticus [248]. Similar observations on delayed neurotoxicity in acute Al encephalopathy have been made in animal studies. Mice and rats seem to be very resistant to the effects of Al, but rabbits are particularly sensitive to Al neurotoxicity and develop severe neurological changes, especially if the metal is administered directly into the central nervous system. Rabbits injected intracerebrally or into the cisterna magna with Al chloride developed quadriparesis and generalized epileptic seizures within seven to 20 days after an incubation period during which they seemed completely normal except for the EEG. Most animals died after a few days of seizures by the 15th day of injection [52, 53]. It seems very obvious that several steps are necessary in the process leading to Al encephalopathy and that each step causes a delay in time before symptoms can occur. It could be postulated that, in the final analysis, no one of these mechanisms will emerge as *the* mechanism but that Al neurotoxicity is due to many, if not all, of them acting synergistically [249].

## 5.4 Pathological Mechanisms of Bone Disease

Much clinical and experimental experience has been obtained about the manifestation of bone diseases, especially in renal patients. Many patients with Al-induced bone disease remain asymptomatic. There are two distinct forms of Al bone disease. The most severe form is osteomalacia, with recurrent fractures and resistance to vitamin D therapy. This disease is characterized by an increase of osteoid due to a mineralization defect induced by Al that is localized at a critical site in the bone, i.e., the osteoid calcification front [250]. The adynamic bone disease is another form of Al-related bone disease, characterized by a reduced bone turnover [97]. Al can have a direct negative effect on the bone by deposition at the mineralization front, causing a defective calcification. This is due to the influence of Al on calcium-phosphorus precipitation, crystal formation and crystal growth [251]. There might also be a toxic effect on the proliferation of osteoblasts and on mature osteoblasts with a time- and dose-dependent effect on osteoblast growth and function [143].

Al also suppresses PTH secretion and maybe as well as PTH synthesis, which results in reduced bone formation and increases Al accumulation in the bone [252]. Other studies have found that:

(1) human parathyroid gland/parathyroid cells exhibit transferrin receptors:
(2) Al-transferrin complex is taken up by the parathyroid gland in a dose-dependent manner; and

(3) uptake of Al by transferrin receptor-mediated endocytosis reduces the secretion of PTH but not its synthesis.

These *in vitro* findings allow us to suggest that transferrin receptor-mediated uptake of Al might, besides other factors such as vitamin D, high calcium dialysate or $CaCO_3$ intake, play a role in the development of hypoparathyroidism associated with Al bone disease. The exact mechanism by which Al-transferrin suppresses iPTH secretion remains to be elucidated [253]. Hyperparathyroidism may afford the bone some protection against the toxic effects of Al [17].

### 5.5 Microcytic Anemia

Most patients with Al intoxication develop an erythropoietin-resistant microcytic anemia in the absence of iron deficiency, and this may be a useful early indication of Al toxicity [41, 93, 254, 255]. The chemical similarity between $Fe^{3+}$ and $Al^{3+}$ suggest that both elements will have similar metabolic effects, suggesting that iron and Al compete during erythropoiesis, resulting from a reversible block in heme synthesis due either to a defect in porphyrin synthesis or to impaired iron utilization. It was also suggested that the main mechanisms for Al toxicity in the erythropoietic system are the interference of Al in the uptake and utilization of iron and an interaction of Al with cellular membrane components, affecting not only their structures but also their functions [256].

## 6 Treatment of Aluminum Intoxications

### 6.1 Prevention and Management of Al Intoxication

Because Al toxicity is a serious iatrogen complication, any preventive measures to avoid exposure to Al is incumbent. Unfortunately, with the current treatment practices, there continues to be a risk that patients receive excessive amounts of Al orally or parenterally. If Al compounds cannot be avoided in patients with renal failure, routine Al tests performed every 3–4 months in serum will identify persons at greatest risk (s-Al levels >100 to 150 μg/L) for chronic Al intoxication. Ideally, any patient with s-Al higher than 40 to 50 μg/L should discontinue Al gels and use other substitutes. These patients should also be warned not to ingest these compounds with liquids containing citrate, for example, fruit juices.

Hemodialysis patients in dialysis centers or on home hemodialysis generally will be protected from Al intoxications by the combination of deionization and RO devices. Nevertheless, one should realize that the composition of the dialysate is a combination of "normal" drinking water according to WHO guidelines and extended water purification. Drinking water conditions can change unexpectedly and erratically at the water plant as well as in the water distribution system, making it absolutely unsuitable for preparation of dialysate upon its arrival in the dialysis unit without optimal further water treatment. Especially Al concentrations in the water supply in regions that use Al salts for purification can vary from day to day and it is impossible to monitor the water Al levels frequently enough to detect periodic but dangerous increments in Al levels.

Physicians in charge of dialysis centers should be aware of the fact that they are solely responsible for the water quality of the dialysate [63] and that the conditions of the water treatment system can change, especially due to Al compounds in the water supply. Membrane fouling in reverse osmosis (RO) systems is inevitable in many systems and effectively anticipating, compensating for and counteracting the fouling phenomena is necessary. While overall flux performances will decline for all membrane systems, the actual time between cleanings will vary, depending upon membrane type, system design and feed water quality. Reverse osmosis is an effective means of removing a wide spectrum of contaminants, including particles, inorganic substances, organic chemicals with molecular weight greater than 300, bacteria and endotoxins. An important exception to this general rule relates to Al which, because of its amphoteric nature, may be present as a non-ionized species. In such circumstances, deionization with mixed bed units may be ineffective for Al removal and reverse osmosis may be superior to deionization. Other factors, especially scalants, can also reduce the quality of the RO membranes. The two most common scalants are calcium carbonate and calcium sulfate. Sharply pointed scale crystals may come into contact and cut the membrane, causing irreversible damage. Well-known fouling particles besides bacteria are Al, iron and silica [259]. Although in a review in the 1980s 92% of reported dialysis dementia cases were dialyzed with either unprocessed or softened water, still 6% of dialysis dementia patients were treated with water prepared by either deionization or reverse osmosis until the 1990s [46]. Even dialysis centers with sophisticated water treatment devices may be faced with episodes of serious Al intoxication, due to the use of Al salt coagulants in potable water treatment. The scarce rainfall in the South of Portugal in 1992 resulted in a subsequent decrease in the level of water sources, resulting in high concentrations of suspended particles which, in turn, necessitated the addition of huge amounts of alum. This action was not reported to the municipal authorities. In a dialysis center in Portugal using a RO, the passage of this severely contaminated water through the water purification installation of a hemodialysis center resulted in the obstruction of the RO membranes [30, 43–45]. The efficiency of the RO membranes relies on two rejection mechanisms: a mechanical sieving filter with 200 dalton pores and an electrostatic repulsive mechanism. The ionized Al fraction is rejected very efficiently (up to 99%) by the RO membranes. However, for the bulk Al, present as a colloid, the RO membranes get fouled. Membrane fouling *does* enhance the accumulation of concentration at the membrane level and masks the electrostatic repulsion mechanism. This is matched with an increased Al breakthrough. In these circumstances, it has been found that the ability of the RO to reject Al may become as low as 30% to 50%. Consequently, RO membranes and cartridge filters had to be replaced frequently. During these interventions, which took place over a period of several days, insufficiently treated water was sent directly to the dialysis machines and to the patients involved. In this case, both ways of exposure probably contributed to acute Al intoxication and all together 18 of 71 (25.3%) patients died of severe encephalopathy [30, 43, 44]. Al levels up to 2200 μg/L were measured in water used for preparation of the dialysate [45]. Most RO units, however, make use of continuous conductivity monitoring of the purified water. A low conductivity represents a low amount of dissolved ions, which means that the membranes are functioning well. Nevertheless, in some cases one series of RO membranes may be insufficient which makes two RO membranes in line necessary [260].

It should be stressed that regular monitoring of the Al levels in serum and dialysate at the normal frequency of one to three times a year [62], might give a false sense of reassurance when the results are normal, as serious exposure might be unnoticed between these testing intervals and serious irreversible neurological damage can develop within weeks. Additional safety precautions can result from a good cooperation between drinking water companies and dialysis centers. Water utilities should notify dialysis centers when situations with important changes in water quality are expected. In particular, dialysis centers should be warned when the Al concentration in drinking water can exceed 30 μg/L. This implies that the dialysis centers should be informed when water utilities change the practice of flocculation to the use of Al flocculants. The centers should also be informed if the concentration of Al exceeds the relatively high level of 200 μg/L, above the WHO standards. In addition, when cemented drinking water distribution pipes are put into service, Al levels should be tested in the tap water and dialysis centers should be warned when Al levels are above 30 μg/L [18].

## 6.2 Treatment

At the start of therapy one should identify and eliminate the source or sources of the Al. Up to date, desferrioxamine is the hallmark of therapy for Al intoxication [260–273].

## 6.3 Desferrioxamine (DFO)

Desferrioxamine (DFO) is a trihydroaminic acid obtained from isolates of *Streptomyces pilosus*. Since 1963 it has been clinically used as an iron-chelating agent in patients with iron overload [261]. DFO effectively chelates trivalent ions such as iron and Al, producing respectively ferrioxamine and aluminoxamine [12, 30, 260–269]. DFO displays rather complicated physicochemical characteristics. Unchelated DFO is a straight chained lipophilic molecule that can penetrate plasma membranes and undergo metabolic breakdown. In contact with Al, it twines itself around the metal to form stable hydrophilic complexes [260, 261]. Since 1980 it is also used in the diagnosis and treatment of Al overload in dialysis patients [11]. DFO treatment has a number of side effects such as digestive dysfunction, skin rashes, neutropenia, hypotension, auditory impairment, cataract, visual impairment, retinal changes similar to those seen in retinitis pigmentosa and life-threatening infections with non-siderophore-producing germs, especially mucormycosis with a fatality rate of 86%. This increased susceptibility to mucormycosis, especially the *Rhizopus* species, is due to the by product of DFO therapy, ferroxamine. Ferroxamine can promote the growth and pathogenicity of specific *Rhizopus* species with receptors for ferroxamine [262]. In addition, exacerbation of Al encephalopathy is an important side effect, which limits high doses of this chelating agent in Al intoxications. The exacerbation of the Al encephalopathy might also be the result of redistribution of Al mobilized by DFO into the brain when very high plasma Al levels in excess of 500 μg/L occur [74]. It has also been speculated that the Al chelated compound, that is, aluminoxamine, by its ability to cross the blood-brain barrier might precipitate or exacerbate Al-related encephalopathy [80, 219]. On the other hand, DFO itself might be a possible causative agent of these neurotoxic effects as the neurotoxicity is dose-dependent and it is also observed in patients with normal renal function [266, 267]. Chelation of Al with DFO has been very useful in the treatment of Al encephalopathy, but the best mode of administration, optimum dose, and duration of treatment are not clearly established. The current practice is to administer DFO once per week; it can be given intramuscularly the evening prior to dialysis, or several hours before dialysis so that sustained ferroxamine levels are present in the plasma for the shortest possible time before being removed. Removal of Al from Al intoxicated patients is difficult due to its large volume of distribution, high protein binding, and poor dialyzability [268]. The Al chelation by DFO is less dose-dependent compared to iron chelation. Apparently there exists a readily available pool of Al, which can be chelated by relatively low doses of DFO, while a deeper pool is more difficult to reach even with higher doses of the chelating agent [269].

## 6.4 Treatment with DFO in Hemodialysis Patients

The optimal dose of DFO for long-term chelation is uncertain, although toxicity appears to be reduced at lower doses. In the search for an optimal dose one should consider the following aspects.

1) Side effects of treatment with DFO.
   a) The potential side effects are, at least in part, dose- and duration-related. With the high doses used in the past, varying between 30 and 80 mg/kg, serious side effects have been reported even after a single dose of the chelator. These include hypotension, exacerbation of Al-related encephalopathy, retinal and auditory neurotoxicity, rash and fatal bacterial and fungal infections. Of 89 patients receiving nightly subcutaneous DFO for transfusion-dependent thalassemia major, 13

presented with visual loss or deafness of acute onset or both. Detailed ophthalmologic, audiologic, and evoked-potential studies uncovered abnormalities caused by neurotoxicity in 27 more. Four patients with visual loss had optic neuropathy, with a marked decrease in acuity, loss of color vision, and delayed visual evoked potentials. Five asymptomatic patients had changes in the pigment of the retinal epithelium. The hearing loss was characterized by a high-frequency sensorineural deficit, which necessitated hearing aids in six patients. When desferroxiamine was stopped, recovery of vision was complete in 2 patients and partial in 2, and in 22 patients with abnormal audiograms, reversal of the hearing deficit was complete in 4 and partial in 1 [264].

b) High DFO treatment dosage requires the need of regular follow-up (at least every three months). Thus; patients treated with DFO should be referred to ophthalmology and otorhinolaryngology for baseline and follow-up examinations [264].
c) The dosage and interval of DFO treatment should be adjusted to keep peak s-Al levels below 400–500 μg/L, to avoid exacerbation of neurologic symptoms [265].
d) Low dose DFO (5 mg/kg) has been recommended at the Consensus Conference on diagnosis and treatment of Al overload in end stage renal failure in 1992 in Paris [30].
e) DFO infusion can cause hypotension. This can be reduced by slowing the infusion rate to one hour, and treated by temporarily stopping the infusion, and administration of a volume expander if required [30].
f) A dosage of 5 mg/kg DFO can differentiate between patients with Al-related bone disease; increased risk for Al toxicity and Al overload. A serum increment after DFO of more 50 μg/L or more had a sensitivity of 91% and a specificity of 95% in the detection of Al overload, thus suitable to detect patients at an increased risk [30].

2) Techniques of treatment with DFO:
   a) The removal of aluminoxamine using conventional dialyzers is modest. However, it can be increased substantially when used in combination with a charcoal hemoperfusion column (ALUCART) or by replacing the conventional dialyzers by high flux polysulfone dialyzers. With the latter devices up to 80% of both the total aluminoxamine and ferrioxamine can be extracted from the body during a single dialysis session. The presence of DFO and the use of a charcoal column (ALUCART) in combination with a conventional dialyzer may markedly increase the extraction of Al during dialysis [273].
   b) Treatment of Al overload using a cartridge with immobilized DFO is efficient and safe and eliminates the potential toxicity of DFO treatment. Although this seems to be the treatment of choice in severely intoxicated patients, this device is not readily available and may be expensive [273].
   c) The high flux polysulfone F-80 dialyzer effects a much greater removal of the DFO-chelated complexes, than conventional dialyzers [272].

3) Timing of DFO treatment and monitoring of Al levels: Administering a test dose of DFO and measuring the increment of serum or plasma Al may estimate the body burden of Al. Patients with Al levels above 300 μg/L at any time are at increased risk for neurotoxicity during the treatment. The treatment schedule should be adjusted in these patients.
   a) High risk patients (peak s-Al levels 300–500 μg/L). Peak aluminoxamine levels occur several hours after administration of DFO and there is no decline of serum aluminoxamine levels during the interdialytic period. Unchelated DFO is no longer detectable 6 h (i.e., 3 half lives) after administration [274]. Especially patients with high Al levels (above 300–500 μg/L) are at risk for developing neurotoxicity symptoms during these first 6 hours. Therefore dialysis should cover these hours in these patients and DFO should be administered i.m. 5 hours prior to a dialysis session with a high flux dialyzer and side effects of DFO infusion will then be minimal [30].
   b) Low risk patients (s-Al level <300 μg/L). Patients with s-Al levels below 300 μg/L are at lower risk of neurotoxicity and the chelator can be given safely during the last hour of a dialysis session at the start of which a serum sample is taken for Al determination. A second serum sample is taken at the start of the next dialysis session.
4) Practical considerations concerning DFO treatment in hemodialysis patients: the treatment schedules not only should avoid complications, but they also should accommodate the patients and dialysis staff as much as possible and therefore fit into the regular dialysis sessions. Considering this, as an example, dialysis should not be necessary during the weekends, high Al levels should be avoided during the weekends, and i.m. injections the day before dialysis should be given only when strictly necessary.
5) Duration of treatment:
   the optimal duration of treatment is not clearly established. In any case one should want to reduce the s-Al levels below normal values (60 μg/L) in dialysis patients, without increment of Al levels above 50 μg/L at two successive occasions after low dose DFO treatment. This will lead to treatment schedules for over a year in some patients, especially those without residual diuresis. On the other hand, Al intoxication is an iatrogenic disorder and maybe therefore one should consider treating patients until their Al burdens have reached the levels before the intoxication (approximately below 20 μg/L) especially when the Al intoxication is due to accidental poisoning (see Table 4).

Considering all the above-mentioned facts, a treatment schedule was developed by W. H. Boer (University of Utrecht, The Netherlands), K. Berend (Diatel Curaçao) after consultation of M. E. de Broe (University of Antwerpen, Belgium). In our experience and that of Barata et al. [30] the dose of 5 mg/kg once or twice a week seems safe enough for long-term treatment and a further decrease of this dosage seems unnecessary, although a recent study demonstrated similar efficiency with half this dosage [265].

## 7
## Conclusions

Al is a very potent neurotoxicant, with a wide variety of pathological, neurochemical, and behavioral consequences of Al exposure. There is no single unifying mechanism that can explain the wide ranges of toxic effects of Al and hundreds of cellular processes both in man and animals have been demonstrated in plants and aquatic animals in nature, experimental animals by several routes of exposure, and under different clinical conditions in humans.

Acute Al toxicity is an iatrogenic pathology that has a asymptomatic phase, whereafter severe epileptic manifestations like myoclonus and seizures result in coma and a high mortality. Especially hemodialysis patients are at risk when the dialysate is contaminated with high Al. The use of Al in bladder irrigation can lead to acute Al encephalopathy in patients with reduced kidney function. Not only kidney patients are at risk for acute Al encephalopathy, also the use of cement in otosurgery has been related to acute Al encephalopathy. Therefore, every effort should be undertaken to limit the use of compounds containing Al in patients. Unfortunately, despite widespread agreement on this point, the abolishment of Al in medicine has only partially been put in practice. Understanding the trends and basics of materials science should help the clinician to avoid unnecessary exposure. However, while Al has no biological function in mankind, it is the most abundant metal in the earth's crust and the widespread occurrence of Al, both in the environment, foodstuffs and medicine, makes it virtually impossible for man to avoid exposure. Up to date, despite an abundance of literature and almost forty years of research, many questions remain unresolved and several aspects of the Al metabolism, as well as the mechanisms of toxicity at the bone and brain level remain to be elucidated.

## 8
## Recommendations for Further Study

1) Adopt a multidisciplinary perspective in Al research, involving collaborative research efforts among scientists from many different specialties.
2) Provide a full overview of the different drugs, clinical materials containing Al.
3) Determine Al concentrations in parenteral fluids that are safe for infants and children and enforce these standards.
4) Provide a better characterization of the brain pathology in dialysis patients, encompassing the entire spectrum from acute and chronic Al encephalopathy in patients with and without clinical signs of neurotoxicity, with emphasis on variations in neuropathology from acute, extremely high-level exposure to long-term, low-level exposure. All existing pathology material should be reevaluated for this purpose.

5) Cement linings of drinking water distribution pipes should be reevaluated for their possible toxicological properties.

**Acknowledgement.** We are indebted to Ella Rijnschot for secretarial help.

## 9 References

1. Wolff FA, Van der Voet GB (1990) In: De Broe ME, Coburn JW (eds), Developments in nephrology. Aluminium and renal failure. Kluwer, Dordrecht, The Netherlands, p 41
2. Stauber JL, Florence TM, Davies CM, Adams MS, Buchanan SJ (1999) J AWWA 91: 84
3. Wilhelm M, de Jager DE, Ohnesorge FK (1990) Pharmacol Toxicol 66: 4
4. Yokel RA, McNamara PJ (2001) Pharmacol Toxicol 88: 159
5. Siem Quoted in Doeilken (1897) Naunyn Schmiedebergs Arch Exp Pathol Pharmak 40
6. Scherp HW, Church CF (1937) Proc Soc Exp Biol Med 36: 851
7. Spofforth J (1921) Lancet I 1301
8. McLauglin AIG, Kazantis G, King E, Teare D, Porter RJ, Owen R (1962) J Indust Med 19: 254
9. Liprese J (1975) Comptes Rendus Seances Societe Biologie 169: 282
10. Alfrey AC, Mishell JM, Burks J, Contiguglia SR, Rudolph H, Lewin E, Holmes JH (1972) Trans Am Soc Artif Int Organs 18: 257
11. Ackrill P, Ralston AJ, Day JP, Hodge KC (1980) Lancet 2: 692
12. Arze RS, Parkinson IS, Cartlidge NE (1981) Lancet 1116
13. Altmann P (1993) Nephrol Dial Transpl (Suppl 1): 25
14. Altmann P, Plowman D, Marsh F, Cunningham J (1988) Lancet 7: 1012
15. Alfrey AC (1980) Neurotoxicology 1: 43
16. Alfrey AC (1986) Kidney Int 29 (Suppl 18): S-53
17. Berend K, Van Der Voet G, Boer WH (2001) Kidney Int 59: 746
18. Berend K, Trouwborst T (1999) J AWWA 91: 91
19. Berkseth RO, Shapiro FL (1980) In: Schneider GE, Winchester JF (eds). Controversies in Nephrology. Georgetown University Press, p 42
20. Altmann P (2001) In: Exley C (ed), Aluminum and Alzheimer's disease. The science that describes the link. Elsevier Science B.V. Amsterdam, p 1
21. Bugiani O, Ghetti B (1982) Neurobiol Aging 3: 209
22. Bugiani O, Ghetti B (1990) In: De Broe ME, Coburn JW (eds), Developments in nephrology. Aluminium and renal failure. Kluwer, Dordrecht, The Netherlands, p 109
23. Van Landeghem GF (1998) Thesis, Leiden, The Netherlands
24. Davison AM, Walker GS, Oli H, Lewins AM (1982) Lancet 785
25. Dewberry FL, Mc Kinney TD, Stone WJ (1980) ASIAIO J 3: 102
26. Elliott HL, Dryburgh F, Fell GS (1978) Br Med J 1101
27. Elliott HL, Mcdoughall AI, Fell GS, Gardiner PHE, Williams ED (1980) Dial Transpl 9:1027
28. Flendrig JA, Kruis H, Das HA (1976) Lancet 1235
29. Flendrig JA (1976) Proc EDTA 13: 355
30. Barata JD, D'Haese PC, Pires C, Lamberts LV, Simoes J, De Broe ME (1996) Nephrol Dial Transplant 11: 125
31. Ganrot PO (1986) Environ Health Perspect 65: 363
32. Kaehny WD, Alfrey AC, Holman RE, Shorr WJ (1977) Kidney Int 12: 361
33. Nadel AM, Wilson WP (1976) Neurology 26: 1130
34. O'Hare JA, Callaghan NM, Murnaghan DJ (1983) Medicine 62: 129
35. Parkinson IS, Ward MK, Feest TG (1979) Lancet 406
36. Platts MM, Anastassiades E (1981) Clin Nephrol 15: 223

37. Rozas VV, Port FK, Fasterling RF (1978) J Dialysis 2: 459
38. Rozas VV, Port FK, Rutt WM (1978) Arch Intern Med 138: 1375
39. Rurwen DR, Olsen SM, Rland IA, Arduino MJ, Reid MH, Jarvis WR (1995) Kidney Int 48: 469
40. Schreeder MT, Favero MS, Hughes JR, Petersen NJ, Bennett PH, Maynard JE (1983) J Chron Dis 36: 581
41. Wills MR, Savory J (1983) Lancet 2: 29
42. Sideman S, Manor D (1982) Nephron 31: 1
43. Stragier A (1994) Nephrol News & Issues 5: 5
44. Simoes J, Barata JD, Haese, de Broe ME (1994) Nephrol Dial Transplant 9: 67
45. Simoes J, Barata JD (1994) Abstract EDTA, Vienna 1002
46. Wing C, Brunner H, Brynger, Chatler C, Donckerwolocke BA, Gurland HJ, Jacobs C, Kramer P, Selwood NH (1980) Lancet ii: 190
47. Farnell BJ, Deboni U, Crapper DR, McLachlan DR (1982) Exp Neuro 78: 241
48. Crapper DR, Dalton AJ (1973) Physiol Behav 10: 935
49. De Boni U, Otvos A, Scott JW, Crapper DR (1976) Acta Neuropathol 35: 285
50. Ghetti B, Gambetti P (1983) Brain Res 277: 388
51. Reusche E (1997) Acta Neuropathol 94: 612
52. Wisniewski HM, Terry RD (1970) In: Wolstonholme GEW, Oconnor M (eds), Ciba foundations Symposium on Alzheimer's disease and related conditions, Churchill, London, p 223
53. Yates CM, Gorden A, Wilson H (1976) Neuropathol Appl Neurobiol 2: 131
54. Huang Y, Herman HM, Lui J, Kattsetos CD, Wills MR, Savory J (1997) Brain Res 771: 213
55. Zatta PF (1995) Med Hypotheses 44: 169
56. Savory J, Huang Y, Wills MR, Herman MM (1966) (abstract) Soc Neurosci 22: 974
57. Wisniewski HM, Struman JA, Shek JW (1980) Ann Neurol 8: 479
58. D'Hease PC, Shaheen FA, Huraib SO (1995) Nephrol Dial Transplant 10: 1 838
59. Edwardson JA, Moore PA, Ferrier IN, Lilley JS, Newton GWA, Barker J, Templar J, Day JP (1993) Lancet 342: 211
60. Fasman GD, Moore CD (1994) Proc Natl Acad Sci USA 91: 11232
61. Flaten TP (2001) Brain Res Bull 55: 187
62. Bakir AA, Hryhorczuk DO, Berman E, Dunea G (1986) ASAIO Trans 12: 171
63. Association for the Advancement of Medical Instrumentation (1993) Water quality for hemodialysis, 2nd edn. Arlington Press, Arlington, VA
64. Freeman RM, Lawton RL, Chamberlain MA (1967) N Engl J Med 276: 1113
65. Van Ginkel MF, van der Voet GB, de Wolff FA (1990) Clin Chem 36: 658
66. Savory J, Garruto RM (1998) Nutrition 14: 313
67. Altmann P, Butter KC, Plowman D, Chaput de Saintonge DM, Cunnigham J, Marsh FP (1987) Kidney Int 32: 710
68. Morris AB (1997) Curaçao, Kompania di Awa I Elektrisidat di Korsou, N.V. (K.A.E). Desalination & Water Reuse 8: 40
69. Douglas BD, Merill DT, Catlin JO (1996) J AWWA 88: 99
70. Brun A, Dictor M (1981) Acta Pathol Microbiol Scand 89: 193
71. Deavidsen AM, Walker GS, Oli H (1982) Lancet 785
72. Kahlil-Manesh F, Agness C, Gonick HC (1989) Nephron 52: 329
73. Madison DP, Baehr ET, Bazell M, Hartman RW, Mahurkar SD, Dunea G (1977) J Speech Hearing Dis 42: 238
74. Sherrard DJ, Walker JV, Boykin JL (1988) AM J Kidney Dis 29: 261
75. Boyce BF, Fell GS, Elder HY, Junor BJ, Elliot HL, Beastall G, Fogelman I, Boyle IT (1982) Lancet 2: 1009
76. Mion JC, Branger B, Issautier R, Ellis HA, Rodier M, Shaldon S (1981) Trans Am Soc Artif Intern Organs 27: 634
77. Masselot JP, Adhemar JP, Jandon MC, Klienknecht D, Galli A (1978) Lancet 2: 1386
78. Nathan E, Pederson S (1980) Acta Pedriatr Scand 69: 793

79. Delanty ND, Vaughan CJ, French JA (1998) Lancet 352: 383
80. Van Lanheghem GF, D'Haese PC, Lamberts LV, Barata JD, de Broe ME (1997) Nephrol Dial Transplant 12: 1692
81. Cumming AD, Simson G, Bell D, Lowie J, Winny RJ (1982) Lancet 1: 103
82. Lowermoor Incident Health Advisory Group (1989) Water Pollution at Lowermoor, North Cornwall. Truro: Cornwall and Isles of Scilly Health Authority
83. Kerr DNS, Ward MK, Ellis HA, Simpson W, Parkinson IS (1992) In: Wiley (ed), Aluminum intoxication in renal disease. Chichester, Ciba Foundation Symposium 169: 123
84. Day JP, Barker J, Evans IJA, Perks J, Seabright PJ, Ackrill P, Lilley JS, Drumm PV, Newton GWA (1991) Lancet 337: 1345
85. Exley C, Burgress E, Day JP, Jeffery EH, Melethil S, Yokel RA (2001). In: Yokel RA, Golub MS (eds), Research issues in aluminum toxicity. Taylor & Francis. Washington, p 118
86. Drüeke TB, Jouhanneau P, Banide H, Lancour B, Yiou F, Raisbeck G (1997) Clinical Science (Colch) 92: 63
87. Jouhanneau P, Lacour B, Raisbeck GM, Yiou F, Banide H, Brown ET, Drueke TB (1993) Clin Nephrol 40: 244
88. Mcmillan TN, Freemont TM, Herxheimeer A, Denton J, Taylor AP, Paxianas M (1993) Camelford Water Poisoning Accident: Serial Neurophysiological Assessments and Further Observations on Bone Aluminum. Hum Exp Toxicol 12: 37
89. Eastwood JB, Levin GE, Pazianas M, Taylor AP, Denton J, Freemont AJ (1990) Lancet 336: 462
90. Nordal KP, Dahl E, Halse J, Sodal G, Thomassen Y, Aaseth J (1986) Acta Pharmacol Toxicol 59 (Suppl 1)
91. Nordal KP, Dahl E, Halse J, Aksnes L, Thomassen Y, Flatmark A (1992) J Clin Endocrinol Metab 74: 1140
92. Reusche E, Koch V, Friedrich HJ, Nunninghoff D, Stein P, Rob PM (1996) Clin Neuropathol 15: 342
93. Touam M, Martinez F, Lacour B, Bourdon R, Zingraff J, Di Giulio S, Drueke T (1983) Clin Nephrol 19: 295
94. Julka D, Vasishta RK, Gill KD (1996) Biol Trace Elem Res 52: 181
95. Short AIK, Winney RJ, Robson JS (1980) Proc Eur Dail Transpl Assoc 17: 226
96. Hdez-Jaras J, Galan A, Sanchez P (1998) Nephron 78: 274
97. Drueke T (1993) Nephrol Dial Transpl 1: 12
98. Mahieu S, del Carmen Contini M, Gonzalez M, Millen N, Elias MM (2000) Toxicol Lett 111: 235
99. Adhemar JP, Laederich J, Jandon MC, Masselot JP, Buission C, Galli A, Klienknecht D (1980) EDTA Proc 17: 234
100. Hughes J, Schreeder MT (1980) EEG in dialysis encephalopathy. Neurology 11: 1148–54
101. Spehr W, Sartorius H, Berglund K, Hjorth B, Kablitz C, Plog U, Wiedemann PH, Zapf K (1977) Elec Clin Neurophy 43: 787
102. Franceschetti S, Bugiani O, de Curtis M, Tagliavini F, Spreafico R, Avanzini G (1986) (Abstract) Neurosci Lett (Suppl 26): S537
103. Winkelman MD, Ricanati ES (1986) Hum Pathol 17: 823
104. Bugiani O, Ghetti B (1988) In: Broggi G (ed), The rational basis of the surgery of epilepsy. John Libbey, London, Paris, p 143
105. Vecchierini-Blineau MF, Thebaud HF, Rrochard D, Coville P (1980) Nephrology 1: 29
106. Levine SN, Sonnier GB, Abreo K (1990) Toxicology 65: 137
107. Royer RJ, Delongeas JL, Netter P, Faure G, Mur JM, Burnel D, Gaucher A (1982) Pathol Biol 30: 211
108. Netter P, Kessler M, Gaucher A, Burnel D, Fener P (1988) Lancet 886
109. Andress DL, Kopp JB, Maloney NA, Coburn JW, Sherrard DJ (1987) N Engl J Med 5: 292–296
110. Vincenti F, Arnaud SB, Recker R, Genant H, Amend WJ Jr, Feduska NJ, Salvatierra O Jr (1984) Kidney Int 25: 677

111. Van der Voet GB, de Haas EJ, de Wolff FA (1985) J Anal Toxicol 9: 97
112. Renard JL, Felten D, Bequet D (1994) Lancet 344: 63
113. Reusche E, Rohwer J, Forth W, Helms J, Geyer G (1995) Lancet 345: 1633
114. Geyer G, Baier G, Helms J (1998) J Laryngol Otol 112: 344
115. Granstrom G, Holmquist J, Tjellstrom A (2000) Ear Nose Throat J 79: 495
116. Hantson P, Mahieu P, Gersdorff M, Sindic CJ, Lauwerys R (1994) Lancet 344: 1647
117. Morell MA, Rajanna S (2001) In: Massry SG, Glassock RJ (eds), Textbook of nephrology. Fourth edition. Lippincott Williams & Wilkins. Wolters Kluwer, Philidelphia, p 1505
118. Daugirdas JT, Blake PG, Ing TS (eds) (2001) Handbook of Dialysis, 3rd edn. Lippincott Williams & Wilkins. Wolters Kluwer, Philidelphia, PA
119. Miller RG, Kopfler FC, Ketty KC, Stober JA, Ulmer NS (1984) J AWWA 76: 84
120. Lusardi PJ, Consonery PJ (1999) J AWWA 91: 28
121. National Secondary Drinking Water Regulations (1991) Final Rule. Fed Reg 56: 20: 3526
122. World Health Organization (1997) Rolling Revision of WHO Guidelines for Drinking Water Quality. Report of Working Group Meeting on Chemical Substances for the Updating of WHO Guidelines for Drinking-water Quality. WHO EOS/97.7, Geneva
123. Jaqmin H, Commenges D, Letteneur L, Barberger-Gatreau P, Dartrigues JF (1994) Am J Epidemiol 139: 48
124. Reiber SH, Kuhkull WA (1996) AWWA Research Foundation
125. Ackley DC, Yokel RA (1998) Toxicology 15: 127: 59
126. Frazão JM, Coburn JW (2001) In: Massry SG, Glassock RJ (eds), Textbook of nephrology. Lippincott Williams & Wilkins. Wolters Kluwer, Philidelphia, p 1244
127. Kerr DN, Ward MK, Arze RS, Ramos JM, Grekas D, Parkinson IS, Ellis HA, Owen JP, Simpson W, Dewar J, et al. (1986) Kidney Int 29: 58
128. Van Landeghem GF, D'Haese PC, Lamberts LV, Djukanovic L, Pejanovic S, Goodman WG, De Broe ME (1998) Clin Nephrol 50: 69
129. D'Haese PC (1988) Ph D thesis. Antwerpen
130. Ball JH, Butkus DE, Madison DS (1977) Nephron 18: 151
131. Cannata JB, Briggs JD, Junor BJR, Fells GS, Beastall G (1983) Lancet 1: 501
132. Van de Vyver FL, Visser WJ, D'Haese PC, Silva FJ, Thomas H, De Broe ME (1989) Nephrol Dial Transplant 4: 555
133. Curtis JR, Sampson B (1989) Int J Artif Organs 12: 683
134. Shapiro WB, Schilb TP, Porush JG (1984) Trans Am Soc Artif Intern Organs 30: 342
135. Shapiro WB, Schilb TP, Waltrous CL, Levy SR, Porush JG (1983) Kidney Int 23: 536
136. Winship KA (1992) Toxicol Rev 11: 123
137. Salusky I, Foley J, Nelson P, Goodman W (1991) New Eng J Med 324: 527
138. Andreoli SP, Bergstein JM, Sherrard DJ (1984) N Engl J Med 310: 1079
139. Brahm M (1986) Clin Nephr 25: 231
140. Yokel RA (1987) J Trace Elem Electrolytes Health Dis 1: 69
141. Schaefer K (1993) Nephrol Dial Transpl 1: 35
142. Alfrey AC (1993) Ther-Drug-Monit 15: 6
143. Goodman WG (1990) In: De Broe ME, Coburn JW (eds), Developments in nephrology. Aluminium and renal failure. Kluwer Academic Publishers, Dordrecht, The Netherlands
144. Griswold WR, Reznik V, Mendoza SA, Trauner D, Alfrey AC (1983) Pediatrics 71: 56
145. Salusky IB, Coburn JW, Paunier L, Sherrard DJ, Fine RN (1984) J Pediatr 105: 717
146. Sedman AB, Miller NL, Warady BA, Lum GM, Alfrey AC (1984 ) Kidney Int 26: 201
147. Van Ginkel MF, van der Voet GB, van Eijk HG, de Wolff FA (1990) J Clin Chem Clin Biochem 28: 459
148. Sheikh MS, Ramirez A, Emmett M, Santa Ana C, Schiller LR, Fordtran JS (1988) J Clin Invest 81: 126
149. Murphy CP, Cox RL, Harden EA, Stevens DA, Heye MM, Herzig RH (1996) Dtsch Med Wochenschr 121: 777
150. Perazella M, Brown E (1993) Am J Kidney Dis 21: 44
151. Skalsky HL, Carchman RA (1983) J Am Coll Toxicol 2: 405
152. Takashi M, Kondo A, Kato K, Murase T, Miyake K (1988) Urol Int 43: 286

153. Torrecilla C, Aguilo F, Munoz J, De la Pena MD, Serrallach N (1987) Actas Urol Esp 11: 457
154. Seear MD, Dimmick JE, Rogers PC (1986) J Urol 136: 665
155. Phelps KR, Naylor K, Brien TP, Wilbur H, Haqqie SS (1999) Am J Med Sci 318: 181
156. Kanwar VS, Jenkins JJ 3rd, Mandrell BN, Furman WL (1996) Med Pediatr Oncol 27: 64
157. Kavoussi LR, Gelstein LD, Andriole GL (1992) Bone Marrow Transplant 10: 383
158. Reusche E, Pilz P, Oberascher G, Lindner B, Egensperger R, Gloeckner K, Trinka E, Iglseder B (2001) Hum Pathol 32: 1136
159. Hantson P, Mahieu P, Gersdorff M, Sindic C, Lauwerys R (1995) J Toxicol Clin Toxicol 33: 645
160. Leveque C, Soulie D, Sarrazin JL, Hor F, Desgeorges M, Cordoliani YS (1996) J Neuroradiol 23: 168
161. Mion C (1985) Clin Nephr 24 (Suppl 1): S88
162. Sedman AB, Wilkening GN, Warady BA, Lum GM, Alfrey AC (1984) J Pediatr 105: 836
163. Sedman AB, Miller NL, Warady BA, Lum GM, Alfrey AC (1984) Kidney Int 26: 201
164. Sedman AB (1992) Pediatr Nephrol 6: 383
165. Freundlich M, Zilleruelo G, Abitbol C, Strauss J, Faugere MC, Malluche HH (1985) Lancet 2: 527
166. Geary DF, Fennell RS, Andriola M, Gudat J, Rodgers BM, Richard GA (1980) J Pedriat 5: 836
167. Randall ME (1983) Lancet I: 1327
168. Flarend R (2001) In: Exley C (ed), Aluminum and Alzheimer's disease. The science that describes the link. Elsevier Science B.V., Amsterdam p 75
169. Bougle D, Bureau F, Voirin J, Neuville D, Duhamel JF (1992) 16: 157
170. McGraw M, Bishop N, Jameson R, Robinson MJ, O'Hara M, Hewitt CD, Day JP (1986) Lancet I: 157
171. Coni E, Bellomonte G, Caroli S (1993) J Trace Elements Electrolytes Health Disease 7: 83
172. Fernandez-Lorenzo JR, Cocho JA, Rey-Goldar ML, Couce M, Fraga JM (1999) J Pediatric Gastroenterol Nutrition 28: 270
173. Bishop NJ, Morley R, Chir MB., Day JP, Lucas A (1997) N Engl J Med 336: 1557
174. Lione A (1983) Food Chem Toxicol 21: 103
175. Exley C (ed) (2001) Aluminum and Alzheimer's disease. The science that describes the link. Elsevier Science B.V. Amsterdam
176. Drueke TB, Jouhanneau P, Banide H, Lacour B, Yiou F, Raisbeck G (1997) Clin Sci 92: 63
177. Riihimaki V, Hänninen H, Akila T, Kovala T, Kuosama E, Paakkulainen H, Valkonen S, Engström B (2000) Scand J Work Environm Health 26: 118
178. Wulf RJ (1999) In: Laden K (ed), Antiperspirants and deodorants. Marcel Dekker, New York, p 215
179. Schwarze HP, Giordano-Labadie F, Loche F, Gorguet MB, Bazex J (2000) J Am Acad Dermatol 42: 888
180. Flarend R, Bin T, Elmore D, Hem SL (2001) Food Chem Toxicol 39: 163
181. Baylor NE (2000) Proceedings from the Workshop on aluminum in vaccines, May 11–12, 2000, San Juan, Puerto Rico. Sponsored by the National Vaccine Program Office, Centers for Disease Control and Prevention
182. Workshop on aluminum in vaccines, May 11–12, 2000 San Juan Puerto Rico. Sponsered by the national vaccine program office. Centers for Disease Control and Prevention
183. Nieboer E, Gibson BL, Oxman AD, Kramer JR (1995) Environ Rev 3: 29
184. Shen R, Ma JF (2001) J Exp Bot 52: 1683
185. Xie ZM, Ye ZH, Wong MH (2001) Environ Int 26: 341
186. Lione A (1983) The prophylactic reduction of aluminum intake. Food Chem Toxicol 21: 103–109
187. Pennington JA, Schoen SA (1995) In: Gitelman HJ (ed), Aluminum and health: a critical review. Marcel Dekker, New York, p 67

188. Lukiw WJ (2001) In: Exley C (ed), Aluminum and Alzheimer's disease. The science that describes the link. Elsevier Science B.V. Amsterdam 95, p 147
189. Martin RB (1986) Clin Chem 32: 1797
190. Martin BR (1990) In: De Broe ME, Coburn JW (eds), Developments in nephrology. Aluminium and renal failure. Kluwer Academic Publishers, Dordrecht, The Netherlands, p 7
191. Martin RB (1991) J Inorg Biochem 44: 141
192. Lind CJ, Hem JD (1975) US Geological Survey Water Supply Paper 1827-G. Government Printing Office. Washington, D.C.
193. Harris WR, Berton G, Day JP, Exley C, Flaten TP, Forbes WF, Kiss T, Orvig T, Zatta PF (1996) In: Yokel RA, Golub MS (eds), Research issues in aluminum toxicity. Taylor & Francis. Washington, pp 91–116
194. Skalsky HL, Carchman RA (1983) J Am Coll Toxicol 2: 405
195. Reusche E, Lindner B, Arnholdt H (1994) Virchows Arch 424: 105
196. Alfrey AC, Hegg A, Crasswell P (1980) Am J Clin Nutr 33: 1509
197. Platts MM, Goode GC, Hislop JS (1977) Br Med J 657
198. James A, Murnaghan DJ (1984) Am J Med 77: 229
199. Hodsman AB, Sherrard DJ, Alfrey AC, Ott S, Brickman AS, Miller NL, Maloney NA, Coburn JW (1982) J Clin Endocrinol Metab 54: 539
200. Lote CJ, Wood JA, Freeman MS (1994) In: Nicolini M, Zatta PF, Corain B. Harwood (eds), Aluminum in chemistry, biology and medicine. Harword, Switzerland, Life Chem Reports 11: p 259
201. Nolte E, Beck E, Winklhofer C, Steinhausen C (2001) Hum Exp Toxicol 20: 111
202. Priest ND, Newton RJ, Day JP, Talbot RJ, Warner AJ (1995) Hum Exp Toxicol 14: 287
203. Priest ND, Talbot RJ, Day JP, King SJ, Fitfield LK, Cresswell RG (1996) BioMetals 9: 221
204. McLachlan DRC, Lukiw WJ, Wong L, Bergeron C, Bech-Hansen NT (1988) Brain Res 427: 255
205. Favarrato M, Mizzen CA, Mclachlan DR (1992) J Chrom 576: 271
206. Solomon B (2001) In: Exley C (ed), Aluminum and Alzheimer's disease. The science that describes the link. Elsevier Science B.V. Amsterdam 95
207. Bentley DR (2000) Human Molecular Genetics 9: 2353
208. Frederickson CJ, Suh SW, Silva D, Frederickson CJ, Thompson RB (2000) J Nutrition 130: 1471S
209. Deloncle R, Huguet F, Babin P, Fernandez B, Quellard N, Guillard O (1999) Toxicol Lett 11: 65
210. Deloncle R, Guillard O, Huguet F, Clanet F (1995) Biol Trace Elem Res 47: 227
211. Somova LI, Missankov A, Khan MS (1997) Methods Find Exp Clin Pharmacol 19: 599
212. Reusche E, Koch V, Friedrich HJ, Nunninghoff D, Stein P, Rob PM (1996) Clin Neuropathol 15: 342
213. Reusche E, Seydel U (1993) Acta Neuropathol 86: 249
214. D'Haese PC, Couttenye MM, De Broe ME (1996) Nephrol Dial Transplant 11 (Suppl 3): 74
215. Banks WA, Kastin AJ, Fasold MB (1988) J Pharmacol Exp Therapeut 244: 579
216. Yokel RA, Lidums V, McNamara PJ, Ungerstedt U (1991) Toxicol Appl Pharmacol 107: 153
217. Lukiw WJ, Bazan NG (2000) Neurochem Res 25: 1173
218. Walton J, Tuniz C, Fink D, Jacobsen C, Wilcox D (1995) Neurotoxicology 16: 187
219. Roskams AJ, Connor JR (1990) Proc Natl Acad Sci USA 87: 9024
220. Exley C ( 1999) J Inorg Biochem 76: 133
221. Yokel RA, Allen DD, Ackley DC (1999) J Inorg Biochem 76: 127
222. Yokel RA, Rhineheimer SS, Brauer RD, Sharma P, Elmore D, McNamara PJ (2000) Toxicol Sci 54: 35
223. Yokel RA, Rhineheimer SS, Sharma P, Elmore D, McNamara PJ (2001) Toxicol Sci 64: 77
224. Van Ginkel MF, van der Voet GB, Marini E, van Keep JP, de Wolff FA (1989) J Neurol Sci 93: 157
225. Dwork AJ, Schon EA, Herbert J (1988) Neuroscience 27: 333

226. Van Landeghem GF, D'Haese PC, Lamberts LV, De Broe ME (1994) Anal Chem 15: 216
227. Connor JR, Menzies SL, St. Martin SM, Mufson EJ (1992) J Neurosci Res 31: 75
228. Guthrie PB, Knappenberger J, Segal M, Bennett MV, Charles AC, Kater SB (1999) J Neurosci 15: 520
229. Neill D, Leake A, Hughes D, Keith AB, Taylor GA, Allsop D, Rima BK, Morris C, Candy JM, Edwardson JA (1996) J Neurosci Res 46: 395
230. Levesque L, Fraser PE, McLachlan DR (1997) J Neurochem 69: S43
231. Sass JB, Ang LC, Juurlink BHJ (1993) Brain Res 621: 207
232. Van Rensburg SJ, Daniels WMU, Potocnik, van Zyl JM, Taljaard JJF, Emsley RA (1997) S Afr Med J 87: 1111
233. Barnard SA, Simon J, Webb TE (1997) Mol Neurobiol 15: 103
234. Di Virgilio F, Chiozzi P, Falzoni S, Ferrari D, Sanz JM, Cytolytic V (1998) Cell Death Differentiation 5: 191
235. Zinchuk VS, Okada T, Kobayashi T, Seguchi H (1992) Brain Res 815
236. Altmann P, Al-Salihi F, Butter K, Cutler P, Blair J, Leeming R, Cunningham J, Marsh F (1987) N Engl J Med: 327: 80
237. Exley C (1999) J Inorg Biochem 76: 133
238. Siegel N, Hang A (1983) Biochim Biophis Acta 744: 36
239. Lee SJ, Stull JT (1998) J Biol Chem 273: 27430
240. Calton JL, Kang MH, Wilson WA, Moore SD (2000) J Neurophysiol 83: 685
241. Antonov SM, Johnson JW (1999) Proc Natl Acad Sci USA 96: 14571
242. Abou-Seif MA (1997) Ann Clin Biochem 34: 645
243. Trombley PQ (1998) J Neurophysiol 80: 755
244. Franceschetti S, Bugiani O, de Curtis M, Tagliavini F, Spreafico R, Avanzini G (1986) (Abstract) Neurosci Lett (Suppl 26): S537
245. Marrs TC, Maynard RL (1994) In: Vinken PJ, Bruyn GE, Wolff FA (eds), Handbook of clinical neurology. 64 revised series 20. Intoxications of the nervous system part I (1994) Elsevier Science B.V. Amsterdam, The Netherlands, Chapter 15, p 224
246. Slikkerveer A, Wolff FA (1994) In: Vinken PJ, Bruyn GE, Wolff FA (eds), Handbook of clinical neurology. 64 revised series 20. Intoxications of the nervous system part I., Elsevier Science B.V. Amsterdam, The Netherlands, Chapter 23, p 332
247. Marsh DO (1994) In: Vinken PJ, Bruyn GE, Wolff FA (eds), Handbook of clinical neurology. 64 revised series 20. Intoxications of the nervous system part I, Elsevier Science B.V. Amsterdam, The Netherlands, Chapter 26, p 413
248. Foncin JF, El Hachimi KH (1986) In: Bès J (ed), Senile dementias: Early detection. Eurotext, London, 191
249. Banks WA, Kastin AJ (1985) Neuropharmacology 24: 407
250. Margerum DW, Cayley GR, Weatherburn DC, Pagenkopf GK (1996) In: Martell AE (ed), Coordination Chemistry, Vol 2, ACS monograph 174, American Chemical Society, Washington, D.C
251. Christoffersen MR, Thyregod HC, Christoffersen J (1987) Clacif Tissue Int 41: 27
252. Diaz-Corte C, Fernandez-Martin JL, Barreto S, Gomez C, Fernandez-Coto T, Braga S, Cannata JB (2001) Nephrol Dial Transplant 16: 742
253. Smans KA, D'Haese PC, Van Landeghem GF, Andries LJ, Lamberts LV, Hendy GN, De Broe ME (2000) Nephrol Dial Transplant 15: 1328
254. Nasiadek M, Chmielnicka J, Subdys J (2001) Ecotoxicol Environ Saf 48: 11
255. Caramelo CA, Cannata JB, Rodeles MR, Fernandez Martin JL, Mosquera JR, Monzu B, Outeirino J, Blum G, Andrea C, Lopez Farre AJ, et al. (1995) Kidney Int 47: 164
256. Ganchev T, Dyankov E, Zacharieva R, Pachalieva I, Velikova M, Kavaldjieva B (1998) Acta Physiol Pharmacol Bulg 23: 27
257. Nesse A, Garbossa G (2001) In: Exley C (ed), Aluminum and Alzheimer's disease. The science that describes the link. Elsevier Science B.V. Amsterdam, p 261
258. Cochran M, Coates JH, Elliot DC (1989) In: de Broe ME, Coburn JW (eds), Aluminum in Renal Failure, Dordrecht, Kluwer, p 139

259. Faller KA, Murray P, Livingston A (1999) Manual of water supply practices. M46 Reverse osmosis and nanofiltration. American Water Works Association, Denver, Co
260. Fernandez-Martin JL, Canteros A, Alles A, Massari P, Cannata-Andia J (2000) Am J Med Sci 320: 96
261. Moeschlin S, Schnider U (1963) N Engl J Med 269: 57
262. Janssen MJ, van Boven WP (1996) Pharm World Sci 18: 187
263. Berland Y, Charhon SA, Olmer M, Meunier PJ (1985) Nephron 40: 433
264. Boelaert J, de Locht M (1993) Nephrol Dial Transpl 1: 43
265. Canteros A, Diaz-Corte C, Fernandez-Martin JL, Gago E, Fernandez-Merayo C, Cannata J (1998) Nephrol Dial Transplant 13: 1538
266. Douthat WG, Acuna Aguerre G, Fernandez Martin JL, Mouzo R, Cannata Andia JB (1994) Nephrol Dial Transplant 9: 1431
267. Stummvoll HK, Graf H, Meisinger V (1984) Miner Electrolyte Metab 10: 263
268. Yokel RA, Lidums V, Ungerstedt U (1991) Toxicology 66: 313
269. Olivieri NF, Bunic JR, Chew E (1986) N Engl J Med 314: 869
270. Bentur Y, Koren G, Tesoro A, Carley H, Olivieri N, Freedman MH (1990) Apr Clin Pharmacol Ther 47: 478
271. Molitoris RA, Alfrey AC, Alfrey PS, Millar NI (1988) Kidney Int 34: 98
272. Verpooten GA, D'Haese PC, Boelaert JR, Becaus I, Lamberts LV, De Broe ME (1992) Nephrol Dial Transplant 7: 931
273. Anthone S, Ambrus CM, Kohli R, Min I, Anthone R, Stadler A, Stadler I, Vladutiu A (1995) J Am Soc Nephrol 1995: 1271
274. Shibasaki H (2000) Muscle Nerve 23: 321
275. Hirono N, Mori E, Ishii K, Imamura T, Shimomura T, Tanimukai S, Kazui H, Hashimoto M, Yamashita H, Sasaki M (1998) J Neurol Neurosurg Psychiatry 65: 913

# A New Effect of Aluminum on Iron Metabolism in Mammalian Cells

Satoru Oshiro

Biochemical Genetics, Medical Research Institute, Tokyo Medical and Dental University, Tokyo113-8510, Japan
*E-mail: satoru.bgen@mri.tmd.ac.jp*

There is an increasing number of reports about new effects of aluminum on iron metabolism in mammalian cells. In this review, based on our recent work and that by others, we describe how aluminum disturbs iron homeostasis. The effects of aluminum may help to explain the pathogenesis of neurodegeneration in Alzheimer's disease.

**Keywords:** Aluminum, Iron, Transferrin, Iron regulatory protein, Alzheimer's disease

## List of Abbreviations

| | |
|---|---|
| Al | aluminum |
| Fe | iron |
| Tf-IU | transferrin-independent iron uptake |
| IRP | iron regulatory protein |
| HO | heme oxygenase |
| ROS | reactive oxygen species |
| AD | Alzheimer's disease |
| $\beta$-AP | $\beta$-amyloid protein |
| Tf-R | transferrin receptor |

## 1 Introduction

Aluminum (Al) is the most abundant metal in the environment. Ingestion from air, foods, water and medicine is unavoidable. Although no case has Al been shown to have a definite biological function, it may substitute for chemically similar elements in living cells. Chemically similar ions might have a similar metabolism. Hence, Al competes with a similar metal, iron which plays a critical role in mammalian cells. There are accumulating reports on new effects of Al on iron metabolism in mammalian cells. The present review includes recent topics on the relations between Al, iron metabolism-related protein and iron regulatory proteins as well as neuronal cell death by Al. Advances during the last five to ten years on these issues are described.

## 2 Transport Mechanism of Aluminum in Mammalian Cells

### 2.1 Aluminum Transport by Transferrin-Dependent and -Independent Uptake

Mammalian cells take up iron by three kinds of iron (Fe) uptake systems; the transferrin receptor (TfR)-mediated endocytosis, redox, and Tf-independent iron uptake (Tf-IU) systems [1–7] (Fig. 1).

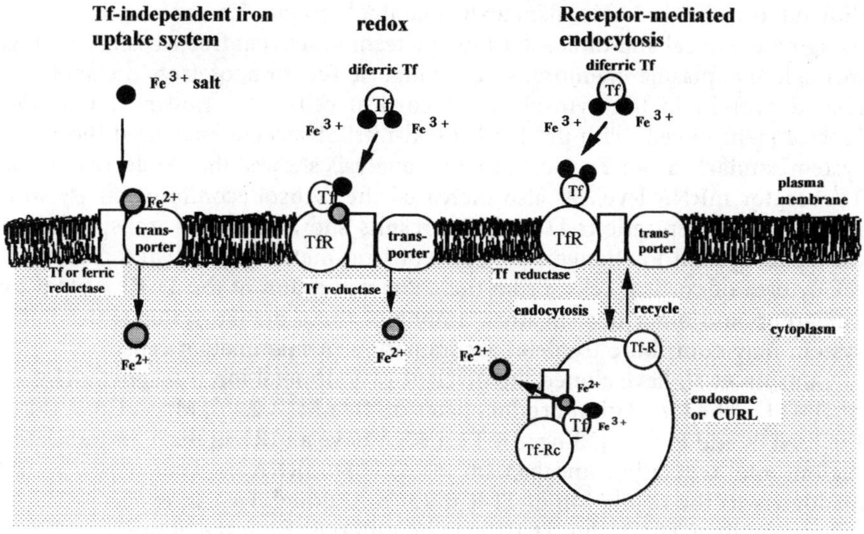


Fig. 1. The three kinds of iron uptake systems in mammalian cells

TfR-mediated endocytosis is a well-known uptake system; Tf binds one or two Fe atoms, but only diferric Tf ($Fe_2Tf$) has a high affinity for TfR to be taken up by the receptor-mediated endocytosis. This system uses a mobilization pathway that involves endosomal acidification, reduction of ferric Fe, and ferrous Fe transport [8]. Recently, it was clarified that divalent cation/ metal ion transporter (DCT1) or Nramp2 involves iron transport from the endosome to the cytosol [9, 10]. Al resembles Fe in chemical characteristics; ionic radius, charge density, and coordination number [11]. Therefore, Al binds with Tf to form di—Al—Tf. Al bound to Tf even passes through the blood-brain barrier to enter the neuronal cells via Tf receptor-mediated endocytosis [12].

The redox system does not depend on endosomal acidification but needs TfR. $Fe_2Tf$ first binds to TfR which is located in close proximity to the proton- and electron-pumping NADH:Tf oxidoreductase. The Fe—Tf bond is destabilized by proton efflux, making $Fe^{3+}$ susceptible to reduction. $Fe^{2+}$ is trapped by a plasma membrane binder and can be transported by a translocator [4]. As Al is a simple trivalent cation incapable of redox changes, it may be theoretically impossible that Al bound to Tf is taken up by a redox mechanism. Actually, no reports on a redox-mediated process of Al bound to Tf have been made.

Although non-Tf-bound Fe uptake occurs in a variety of the cells, detailed molecular mechanisms are unknown at present. However, we made a preliminary experiment on the basic mechanism and found that Tf-independent Fe is reduced at the plasma membrane prior to uptake, as the ferric reduction of $Fe_2Tf$ taken up by the redox mechanism occurs at the same site [5, 13].

Al is taken up not only by receptor-mediated endocytosis but also by the Tf-IU system (Fig. 2). Since Al has no suitable radionuclide, it has been difficult to verify that Al is taken up by mammalian cells. However, it was shown using biochemical and molecular biology techniques that free Al salts can pass through the plasma membrane to change a function of iron metabolism-related protein in the cytosol of rat cortical cells after transport [14, 15]. Pretreatment of cells with μM levels of Al-nitrilotriacetate increased the Tf-IU system similarly as for Fe. Northern blot analysis showed that Al decreased the Tf receptor mRNA level. Al also increased the cytosol aconitase activity with increasing concentrations [14]. These results show that Al is transported by the Tf-IU system to change a function of iron metabolism-related protein in the cytosol of cortical cells. From the characterization of the Tf-IU system by our group and other investigators, among of three uptake systems, the Tf-IU system may contribute to metal accumulation in mammalian cells.

Gunshin et al. have cloned the divalent cation/metal ion transporter (DCT1 or DMT1). The translation product takes up free $Fe^{2+}$, $Zn^{2+}$, $Mn^{2+}$, $Co^{2+}$, $Cd^{2+}$, $Cu^{2+}$, $Ni^{2+}$, and $Pb^{2+}$ [9]. Since DCT1 mRNA have an IRE in the 3′ untranslated region, the regulation mechanism of DCT1 mRNA is considered to be analogous to the regulation of Tf-R mRNA. Although we can expect DCT1 to take up Al by a Tf-IU, this Tf-IU may be different from the above-mentioned Tf-IU because the former is down-regulated by accumulated metals and the latter up-regulated. As summarized in Fig. 2, so far, it has been reported that mammalian cells can taken up Al by Tf-dependent and -independent iron uptake systems.

Although the molecular mechanism of metal uptake via the Tf-IU system is unknown, there is a fair amount of experimental data on the characteristics of

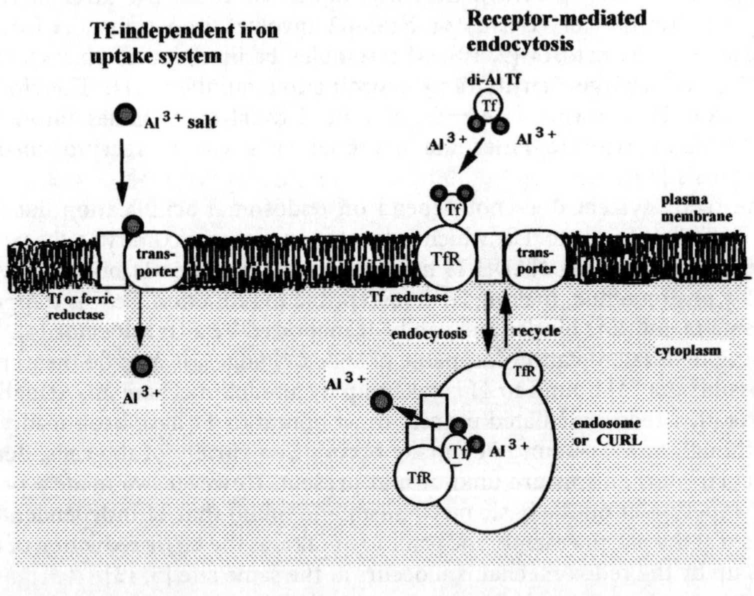


**Fig. 2.** Aluminum uptake by receptor-mediated endocytosis and transferrin-independent iron uptake systems in mammalian cells

multiple Tf-IU systems in various mammalian non-neuronal cells. From their characteristics, these transport systems can be classified into two types: a metal-induced and $Ca^{2+}$-dependent type, e.g., human skin fibroblasts and HeLa cells and a non-metal-induced and $Ca^{2+}$-independent type, e.g., K562 cells [2]. The Tf-IU system of human skin fibroblasts is up-regulated by iron salts. In HeLa cells and fibroblasts, divalent metal ions, such as $Cu^{2+}$, $Zn^{2+}$, and $Fe^{2+}$ inhibit $Fe^{3+}$ uptake. By contrast, the K562 cell transport mechanism is $Ca^{2+}$-independent and not up-regulated by extracellular Fe salts. Uptake of $Fe^{3+}$-NTA by K562 cells was specifically inhibited by $Cd^{2+}$ and $Cu^{2+}$, whereas $Mn^{2+}$, $Co^{2+}$, and $Ni^{2+}$ were ineffective. Quite recently, the presence of a non-Tf-mediated uptake for Fe and Mn in glial cells has been reported [16]. However, the Tf-IU of both neuronal and glial cells has not been characterized.

## 2.2 Characterization of Transferrin-Independent Iron Uptake System in Neuronal and Glial Cells

Therefore, we further examined the characteristics of the Tf-IU system in primary neuronal and glial predominant systems (NP and GP) from rat cerebral cortex thought to be involved in the accumulation of Al and Fe in the brain [17]. The Tf-IU system in cortical cells exhibited properties similar to those of HeLa cells and human skin fibroblasts in two regards: the dependence on $Ca^{2+}$ and the up-regulation by transition metals. Comparing the characteristics in NP and GP cells, there are no differences between them. However, we found that GP cells are more remarkable than NP cells in terms of the degrees of up-regulation of Tf-IU by Al, suppression of Tf-R mRNA expression, and decrease of IRP/IRE complex by Al [15], suggesting that glial cells rather than neuronal cells contribute to the metal accumulation and are more resistant to oxidative stress caused by metals than neuronal cells. The present study may help to explain the pathogenesis of neurodegeneration in AD disorders caused by metal-generated oxidative stress [15].

In relation to our findings that Al up-regulates Tf-IU system in both neuronal glial cells, Abreo et al. observed similar phenomena [18].

They describe the uptake and toxicity of Al, the effect of Al on Fe uptake, and the expression of NFT protein in murine neuroblastoma cells (Neuro 2A). Significant cell Al uptake and inhibition of cell growth were seen in Neuro 2A cells at 1, 2, 3, and 4 days after plating in medium containing Al-Tf and Al citrate. Al-loaded Neuro 2A cells showed increased rates of $^{59}Fe$ and $^{125}I$-Tf uptake and total cellular Fe content at each day after plating compared with control cultures. Significant increases in NFT protein staining were detected in Al-exposed cells at 3 and 4 days in culture compared with controls. The intensity of NFT staining in Al-loaded cells was directly proportional to the time in culture. These results suggest that the accumulation of Al in Neuro 2A cells resulted in increased uptake of Fe, inhibition of cell growth, and expression of NFT protein, partially mimicking the pathological hallmarks of Alzheimer's disease.

# 3
# New Effect of Aluminum on Iron Metabolism in Mammalian Cells

## 3.1
## Effect of Al on Iron Regulatory Protein-1 and -2 in Mammalian Cells

Iron regulatory proteins (IRPs) regulate the cellular iron level in mammalian cells. IRPs are known as cytosol mRNA binding proteins which control the stability or the translation rate of mRNAs of iron metabolism-related proteins such as TfR, ferritin, and 5-aminolevulinic acid synthetase in response to the availability of cellular iron [19–21] after uptake [5]. The regulatory mechanism involves the interaction between the iron-responsive element (IRE) in the 3′ or 5′ untranslated regions of the transcripts and cytosolic IRPs (IRP-1 and -2). IRP-1 is an iron-sulfur (Fe-S) protein with aconitase activity containing a cubane 4Fe-4S cluster. When Fe is replete, IRP-1 prevails in a 4Fe-4S form as a holo-form and is an active cytoplasmic aconitase. As shown in Fig. 3, when Fe is deplete, it readily loses one Fe from the fourth labile Fe in the Fe-S cluster to become a 3Fe-4S cluster and in this state has little enzymatic activity [22, 23].

As it is less known whether IRP-1 modulates various non-Fe metals, we examined whether treatment of homogeneously purified IRP-1 with non-Fe

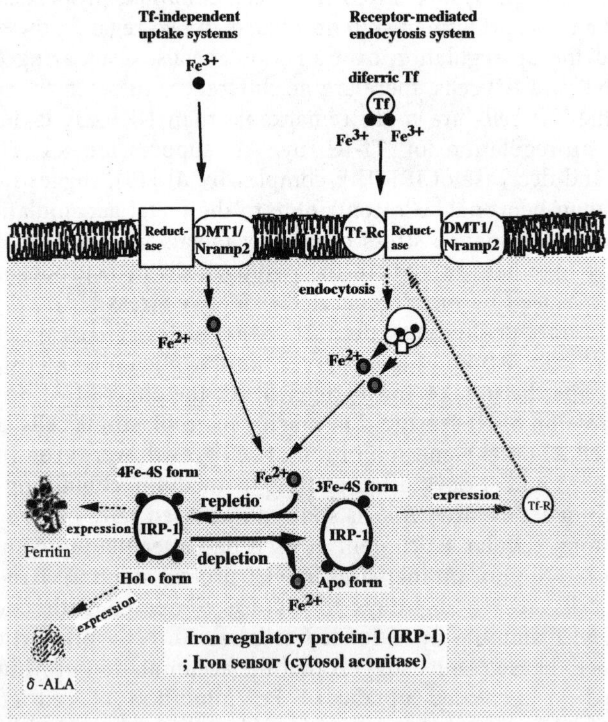


**Fig. 3.** Effect of aluminum on post-transcriptional regulation by iron regulatory protein-1

metals decreased the affinity to IRE in RNA-band shift assays and increased the aconitase activity. As results of these experiments, we showed that the purified IRP-1 directly bind not only Fe, but Mn, Ni, Cu, Cd, and Hg. Also, in competition experiments, non-Fe metals inhibited $^{55}Fe$ incorporation into the fourth labile position of the Fe-S cluster of IRP-1. In PLC hepatoma cells, metal loading inactivated the binding activity and activated the enzyme activity. Non-Fe metals also suppressed TfR mRNA expression in the cells. These results suggest that various non-Fe metals modulate IRP-1 by conversion of a 3Fe-4S apo-form to a [1 non-Fe metal + 3Fe]-4S holo-form [24]. Therefore, as shown in Fig. 4, we present a hypothesis that IRP-1 binds Al to form a [1 Al + 3Fe]-4S holo-form. It is an interesting question whether Al also binds IRP-1 *in vivo* and *in vitro* as Fe does. Actually, we have observed that Al binds to the purified IRP-1 to increase its aconitase activity and decrease its mRNA binding activity (Oshiro S, et al., unpublished data).

In contrast, IRP-2 possesses neither the Fe-S cluster nor aconitase activity [25]. The binding activity of IRP-1 to IRE is modulated by intracellular Fe levels.

With this objective in mind, we examined the effects of Al accumulated in brain cells on the iron metabolism using primary cultures from the fetal rat cerebral cortex. To examine the influence of Al on the iron metabolism, the

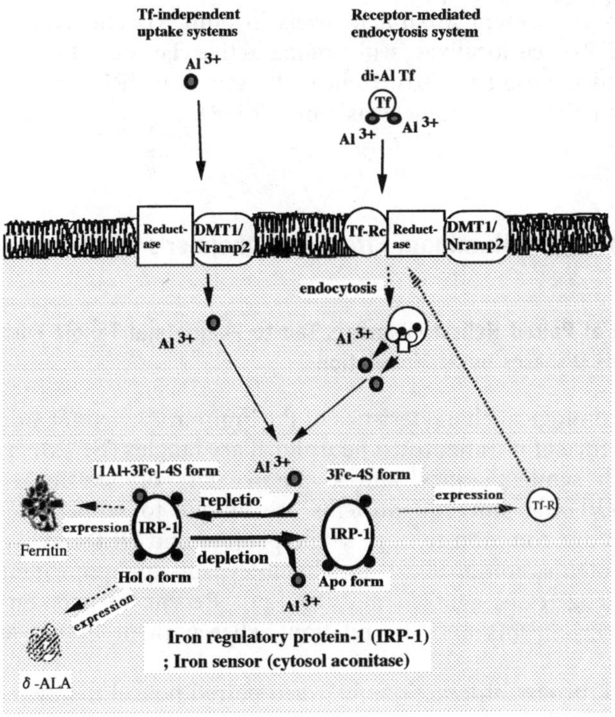


**Fig. 4.** Effect of aluminum on post-transcriptional regulation by iron regulatory protein-1

interaction between Al accumulated in the cells and IRP-1 known as a cellular iron regulator was examined by Northern blot analysis and assay of its aconitase activity; Al decreased the Tf receptor mRNA level and increased the aconitase activity of IRP with increasing concentrations. The increase of the enzyme activities by Al was also observed in an *in vitro* experiment. These results suggest that Al taken up by mammalian cells affects the iron metabolism in the cells. Al seems to interact with IRP to regulate TfR expression.

IRP-2 is also one of the central regulators of iron homeostasis. IRP-2 regulates the expression of molecules involved in iron metabolism by binding to IREs in the transcripts of those molecules in Fe depletion. IRP-2 is degraded efficiently by the proteasome in Fe-replete cells. IRP-2 is oxidized and ubiquitinated in cells before degradation. Moreover, Fe-dependent oxidation converts IRP-2 into a substrate for ubiquitination *in vitro*. Excess Fe is sensed by its ability to catalyze site-specific oxidations in IRP-2, oxidized IRP-2 is ubiquitinated, and ubiquitinated IRP-2 subsequently is degraded by the proteasome [26]. Yamanaka et al. reported that Al antagonizes the Fe-induced decrease in IRE binding activity of IRP-2. Al also inhibits Fe-induced oxidation of IRP2 *in vitro*. These results suggest that Al stabilizes IRP-2 by interfering with the iron-catalyzed oxidation, which results in perturbation of iron metabolism. The action of Al may involve selective targeting and removal of oxidatively modified proteins that may contribute to the turnover of many proteins degraded by the proteasome [27].

The IRP-1 is present at similar levels in both AD and control brains. In contrast, IRP-2 co-localizes with redox-active Fe in NFT, senile plaque neurites, and neuropil threads. Hence, alterations in IRP-2 might be directly linked to impaired iron homeostasis in AD [28].

# 4
# Implication of the Aluminum Effect in Alzheimer's Diseases

## 4.1
## Contribution of Paired Helical Filaments Tau to Plaque and Tangle Formation in Alzheimer's Diseases by Al Modulation

The histopathological characteristics of the brain in Alzheimer's diseases (AD) are the presence of intraneuronal neurofibrillary tangles (NFTs), extraneuronal amyloid-rich senile plaques, and a massive loss of neurons of the telencephalon. Although amyloid proteins are unique to senile plaques, several components are common to both senile plaques and NFTs: hyperphosphorylated tau proteins, ubiquitin, $\alpha$1-antichymotrypsin, apolipoprotein E, heparan sulfate proteoglycans, and Al and Fe [29]. Recent reports on Al and Fe transport mechanisms and the relation between them in AD are reviewed below.

Shin et al. probed interactions between paired helical filaments (PHF) tau, Al salts, and other plaque and tangle components. They investigated the *in vivo* interactions of PHF tau and $AlCl_3$ with other plaque and tangle components by injecting PHF tau with and without $AlCl_3$ into the rodent brain. PHF tau co-injected with $AlCl_3$ formed aggregates that persisted much longer in the rat brain, and induced longer-lived co-deposits of A beta, ubiquitin, $\alpha$1-antichymotrypsin, and ApoE than did PHF tau alone. Injections of PHF tau with $AlCl_3$ also induced neurons near the injection site to acquire PHF tau-like properties as monitored with antibodies that recognize defined PHF tau epitopes containing phosphoserine residues. Injections of $AlCl_3$ alone as well as injections of normal adult and fetal CNS tau, several different synthetic peptides, neurofilament proteins, $\alpha$1-antichymotrypsin, haparan sulfate proteoglycans, or ApoE with and without $AlCl_3$ failed to induce co-deposits of A beta or alter the immunoreactivity of tau in rodent neurons. To determine if Al salts interact directly and specifically with PHF tau *in situ*, they pretreated sections of AD hippocampus with 10 mM $AlCl_3$ and then probed these sections by immunohistochemistry with antibodies to PHF tau as well as to a number of other plaque and tangle components. Preincubation of these sections with $AlCl_3$ diminished PHF tau immunoreactivity in NFTs and senile plaques using the PHF tau-specific antibodies, and PHF1, while the immunoreactivity of other plaque and tangle proteins was not abolished. They also examined the effects of $AlCl_3$ on PHF tau and normal adult human CNS tau *in vitro*. $AlCl_3$ had no effect on normal adult human CNS tau, while increasing concentrations of $AlCl_3$ (from 0.1 to 1.0 mM) induced PHF tau to aggregate at the top of the stacking gel, and at high concentrations (0.3 and 1.0 mM) of $AlCl_3$, PHF tau completely failed to enter the gel. These studies suggest that Al binds to PHF tau, induces these proteins to aggregate, and retards their proteolysis. Further, since intracerebral injections of PHF tau with and without $AlCl_3$ in rats appear uniquely capable of inducing co-deposits of a number of proteins found in authentic AD senile plaques and NFTs, they speculate that the contributions of PHF tau to plaque and tangle formation in AD may be modulated by Al [30].

## 4.2 Effect of Al on Fe Uptake in Neuronal Cells

The uptake and toxicity of Al, the effect of Al on Fe uptake, and the expression of NFT protein in murine neuroblastoma cells (Neuro 2A) were examined in the following study. Significant cell Al uptake and inhibition of cell growth were seen in Neuro 2A cells at 1, 2, 3, and 4 days after plating in medium containing Al-Tf and Al citrate. Al-loaded Neuro 2A cells showed increased rates of $^{59}$Fe and $^{125}$I-Tf uptake and total cellular Fe content at each day after plating compared with control cultures. Significant increases in NFT protein staining were detected in Al-exposed cells at 3 and 4 days in culture compared with controls. The intensity of NFT staining in Al-loaded cells was directly proportional to the time in culture. There was no difference in malonodialdehyde levels measured in control versus Al-loaded Neuro 2A cells. These results suggest that the accumulation of Al in Neuro 2A cells resulted in increased uptake of Fe, inhibition of cell growth, and expression of NFT protein, partially mimicking the pathological hallmarks of AD. This model system may also be applicable for Al-induced dialysis dementia, because the Al concentrations at which cell toxicity occurred can be found in dialysis patients [29]. This report supports our hypothesis [14, 15, 17].

## 4.3 Effect of Al on Fe Metabolism in AD Brain

As mentioned above, there are the profound relationships between Al and Fe metabolism in mammalian cells; Al can bind proteins bound to Fe. Apo-Tf binds to Al to form di-Al-Tf ($Al_2Tf$). $Al_2Tf$ is recognized by TfR to be taken up by brain cells. Al binds Fe storage protein, ferritin and also influences the expression of ferritin mRNA. If this is a reliable phenomenon, Al is required to interact with IRPs which post-trascriptionally regulate the expression of ferritin or TfR mRNAs.

## 4.4 Amount of Al and Fe in Ferritin in AD Brain

One of two groups of rats was fed $AlCl_3$ (100 μM) for 1 year in the drinking water. Brain tissue contained about one-third of the amount of ferritin found in the liver. While brain ferritin from normal rats contained 42.1 ± 14.3 mol of Al, that from the Al-fed group contained 115.4 ± 48.3 mol of Al per mol of ferritin. Liver ferritin from both groups contained similar amounts of both Al and Fe, and the amounts were less than that found associated with brain ferritin. Ferritin isolated from AD brains contained more Al and more Fe than that from age-matched controls [31, 32].

## 4.5 Post-Transcriptional Regulation of Ferritin Heavy Chain in AD Brain

In the aging human brain, the concentrations of iron and its major storage protein, ferritin, rise but the distribution of metal and protein remains non-uniform. More ferritin could be isolated from the brains of humans who died of AD than from age- and sex-matched controls. Also, brain ferritin of rats chronically exposed to $AlCl_3$ in their drinking water contained more Al and Fe. Based on the above earlier observations, a more detailed study of human brain ferritin was initiated. Ferritin was a component of senile plaques in AD. Ferritin obtained from normal or AD brains was composed of 24 subunits [70% heavy (H) chain; 30% light (L) chain]. The techniques of molecular biology revealed the presence of an additional ferritin mRNA species for the H subunit which was more abundant in the brain than in other human tissues. It contained not only the entire sequence of 919 nucleotides of H chain mRNA from liver but also an additional segment of 279 nucleotides in the 3′-untranslated region. The two mRNA seemed to arise by the use of an alternate polyadenylation site of the same primary transcript. Ribonuclease protection assays revealed that the concentrations of the longer mRNA in the normal hippocampus and the hippocampus of patients with AD brains were similar [32]. These results show that expression of ferritin H chain in AD brain does not regulate at the transcriptional level but rather the post-transcriptional level.

## 4.6
## Acceleration of β-AP Generation by Al in AD Brain

Clauberg and Joshi presented an interesting hypothesis that Al accumulated in AD brain accelerates the generation of β-AP [33]. The plaques from AD brain contain β-AP, α1-antichymotrypsin, and Al. They examined the relation of all three components of plaques to each other. The β-AP is derived by proteolytic processing from β-AP and some of these proteins contain a domain that is highly homologous to the bovine pancreatic trypsin inhibitor. Bovine pancreatic trypsin inhibitor also inhibits α-chymotrypsin and they showed that Al affects both the activity and the inhibition of this enzyme. At pH 6.5, in the presence of Al, the enzyme activity is doubled, and the inhibitor is only 1% as effective as in the absence of the metal ion. The inhibition by BX-9, a protease inhibitor prepared from protein components of amyloid plaques, is also reduced by Al; so too is that by α1-antichymotrypsin but to a lesser degree. In the Alzheimer brain, they propose that Al may accelerate proteolytic processing of the α-amyloid precursor protein by suppression of the inhibitor domain. Thus, the β-amyloid protein (β-AP) may accumulate and initiate plaque formation.

## 4.7
## Aggregation and the Deposition of β-AP by Al and Senile Plaques Formation

β-AP, which is the major component of senile plaques in the brains of AD, has an intrinsic tendency to form insoluble aggregates. The aggregation of β-AP enhances its neurotoxicity and is assumed to play a key role in the amyloid deposition. The plaques are predominantly composed of human β-AP β A4, a 40-mer whose neurotoxicity is related to its aggregation. In physiological buffers, $Ca^{2+}$, $Co^{2+}$, $Cu^{2+}$, $Mn^{2+}$, $Mg^{2+}$, $Na^{+}$, or $K^{+}$ at 10 mM had no effect on the rate of β A4 aggregation. In sharp contrast, Al, Fe, and Zn under the same conditions strongly promoted aggregation (rate enhancement of 100- to 1000-fold). The aggregation of β A4 induced by Al and Fe is distinguishable from that induced by Zn in terms of rate, extent, pH-, and temperature-dependence. High concentrations of certain metals may play a role in the pathogenesis of AD by promoting aggregation of beta A4. The aggregation of β-AP (beta 1–40) is promoted by Al [34, 35]. Al concentrations of 15–80 ppm (0.5–3.0 mM) were detected in tangle NFTs of AD [36]. Therefore, it is possible that Al directly influences the process of aggregation and the deposition of senile plaques.

Beta-amyloid precursor proteins (β-APPs) are normal components of the human brain and some other tissues. Proteolysis of these, presumably by serine proteases, generates a 39 to 42 amino acid-long peptide, α-AP. In AD brains, β-AP aggregates into plaque, the hallmark of AD brains. Some of the

$\alpha$-APPs also contain a 56 amino acid-long segment which inhibits serine proteases. *In vitro*, at pH 6.5, Al activates $\beta$-chymotrypsin 2-fold and makes it dramatically resistant to protease inhibitors such as bovine pancreatic trypsin inhibitor (bPTI) or its mimic present in the $\beta$-APPs [30].

## 4.8
## Generation of Reactive Oxygen Species by Al

We observed active oxygen species in rat glial and neuronal cells after treatment of Al- and Fe-NTA using an *in vitro* model of AD brain. Reactive oxygen species (ROS) were measured by using a non-permeant chemiluminescence probe, 2-methyl-6-(*p*-methoxyphenyl)-3,7-dihydroimidazo[1,2-*a*]-pyrazin-3-one hydrochloride (MCLA) specific for superoxide ($O_2^-$) and singlet oxygen ($^1O_2$) [37]. Rat neuronal predominant (NP) or glial predominant (GP) cells were incubated with 50 μM Al- or Fe-NTA at 37 °C for 24 h, then washed three times with PBS. While the cells were being incubated in PBS containing 25 μM MCLA at 37 °C, chemiluminescence was measured for 5 min. In addition, to examine whether $O_2^-$ is produced by NP or GP cells after Al- or Fe-loading, 0.5 μM Cu-Zn superoxide dismutase (SOD) was added to the reaction mixture to examine the production of $O_2^-$ in these cells. Although $FeCl_3$ forms $O_2^-$ in cerebral cortex after intracortical injection it is not known which of the cortical cells generates $O_2^-$. To investigate the mechanism by which the metal toxicity is expressed in these cells, ROS were measured using a chemiluminescence analysis specific for $O_2^-$ and $^1O_2$. In this experiment, we assayed ROS extracellularly released from the cells after metal-loading using MCLA reacting specifically with both ROSs.

For the generation of ROS by Fe-loading, the MCLA-dependent intensity in GP or NP cells was 2.0- or 2.4-times as high as than that in each of the control cells. The intensity in the NP cells with or without Fe treatment increased 3.8- and 1.5- times over that in GP cells without treatment, respectively. The ROS level in Fe-loaded NP cells was 2.0-times that of Fe-loaded GP cells.

For the generation of ROS by Al-loading, the amount of ROS generated by GP or NP cells was 1.5- or 1.8-times than that by each of the control cells. The intensity in the NP cells with or without Al-treatment increased 2.5- and 1.4-times over that in GP cells without treatment, respectively. ROS level in Al-loaded NP cells was 2.5-times higher than that of Al-loaded GP cells. Al-loading generated less ROS than Fe-loading in both cells. In addition, to examine whether $O_2^-$ is produced in these cells, SOD was added to the reaction mixtures containing ROS. In Fe-loaded GP and NP cells, there was a 40 and a 54% decrease of chemiluminescence generated in Fe-loaded cells, respectively. In Al-loaded GP and NP cells, there was a 27 and a 43% decrease compared to GP or NP cells with Al-treatment.

For the generation of $^1O_2$, it is known that $^1O_2$ is involved in oxidative damage to membrane lipids *in vitro*, so it should be detectable [38]. Although we could not clearly show the generation of $^1O_2$ in this experiment, a part of the neuronal cell death we observed might be due to the cell toxicity of ROS. Comparing the sensitivities of the cells to oxidative stress, GP cells were more resistant to Al and Fe toxicities than NP cells. Neurons are more vulnerable than glia to insults such as ischemia, hypoxia, and brain trauma [39]. It was also reported that the antioxidant capacity of neurons in culture is lower than that of glial cells in culture [40]. Our findings are compatible with these reports. The relation between the amount of ROS in both cells and the viabilities of these cells may reflect the difference between the antioxidant defense systems of both cells [15].

## 4.9 Activation of Transcription Factor by Oxidative Stress Generated by Fe in Neuronal Cells

In *in vivo* and *in vitro* studies, Al can potentiate the oxidative stress produced by Fe [41, 42]. Transcription factor, nuclear factor (NF)-$\kappa$B is activated by oxidative stress. The relevance of NF-$\kappa$B to neurodegeneration is suggested by a correlation between the amount of activated transcription factor NF-$\kappa$B and a key inflammatory enzyme, COX-2, in both aging and AD temporal lobe neocortex [43].

## 4.10 Effect of Al and/or Fe on Neuronal and Glial Cell Death

Savory et al. proposed that aging is an important factor in the susceptibility of neurons to oxidative stress and to subsequent apoptosis [44]. They demonstrated that aged rabbits treated intracisternally with Al maltolate exhibit intense intraneuronal silver positivity indicative of the formation of neurofilamentous aggregates, together with oxidative stress. These changes occur in the CA1 region of the hippocampus as well as in cerebral cortical areas. Apoptosis, measured by the TUNEL *in situ* technique, co-localizes with oxidative stress. Young animals treated with Al show few of these alterations, while age-matched controls are essentially negative. Further studies on the time course of these and related changes demonstrate that oxidative stress and redox-active Fe accumulation in hippocampal neurons occur very rapidly, within a period of 3 hours, and increased in intensity at 72 hours. Changes suggestive of apoptosis are seen by 24 hours and are pronounced at 72 hours. In aged animals there is an initially intense immunopositivity at 3 hours for Bcl-2, with negative staining for Bax. By 72 hours, when apoptosis is strongly evident, Bcl-2 is negative and Bax strongly positive. In contrast to the aged rabbits, young animals treated similarly with Al exhibit much less oxidative stress with no apoptosis, and maintain Bcl-2 immunopositivity and negative Bax staining. These findings strongly support the key role that oxidative damage plays in the process of neurodegeneration and in the increased vulnerability to Al-induced injury in the aged animal. These are novel observations which may have important implications for aiding our understanding of the pathogenesis of the neurodegeneration occurring in AD [44].

## 4.11
## Effect of Al on Glial Cell Death

Al exposure and apoptotic cell death have been implicated in several neurodegenerative conditions including AD. For the apoptosis of astrocytes by Al, the proportion of apoptotic cells and cell cycle distribution were determined by flow cytometric analysis. Exposure to Al at low levels of 100 and 200 μM for up to 6 days did not result in apoptosis of the astrocytes. This result is in agreement with our previous result [15]. A dramatic blockage of apoptotic cells was found at 200 μM Al. However, at 400 μM, Al markedly induced the apoptosis of astrocytes, which was associated with a significant change in cell cycle distribution characterized by increase of G2/M phase cells (128%). Measurements of the intracellular $Ca^{2+}$ concentration using the fluorescent calcium indicator dye Fluo-3 demonstrated a significant increase in the levels of intracellular $Ca^{2+}$ after Al treatment. However, no differences were observed among Al-treated groups. These findings suggest that Al induces and blocks selectively the apoptosis of astrocytes, which depends upon the concentrations of Al. Increased intracellular $Ca^{2+}$ may not be the primary mechanism of Al-mediated apoptotic cell death [45].

## 4.12
## Apoptosis Pathway Induced by Al in Brain Cells

Neurodegenerative diseases, including AD are characterized by a progressive and selective loss of neurons. Apoptosis under mitochondrial control has been implicated in this neuronal death process, involving the release of cytochrome c into the cytoplasm and initiation of the apoptosis cascade. However, a growing body of evidence suggests an active role for the endoplasmic reticulum in regulating apoptosis, either independent of mitochondrial, or in concert with mitochondrial-initiated pathways. Members of the Bcl-2 family of proteins have been shown to either inhibit apoptosis, as is the case with Bcl-2, or to promote it, as in the case of Bax. Intracisternal administration of Al-maltolate to New Zealand white rabbits, an animal system relevant to a study of human disease in that it reflects many of the histological and biochemical changes associated with AD, showed that Al-maltolate induces both cytochrome c translocation into brain cytosol and caspase-3 activation. Furthermore, these effects are accompanied by a decrease in Bcl-2 and an increase in Bax reactivity in the endoplasmic reticulum [46].

## 4.13
## Apoptosis of Astrocytes by Al

When the concentration of 1 mM of Al was incubated with cultured astrocytes and neurons, Al accumulated both in neurons and astrocytes, and

Al caused strong changes in the morphology of astrocytes, including shrinkage of cell bodies and retraction of processes after 8–12 days exposure. Exposures over 15–18 days reduced the astrocytes' viability by 50%. The Al-induced degeneration of astrocytes involved the DNA fragmentation characteristic of apoptosis, and staining of Al-treated astrocytes with the DNA-binding fluorochrome Hoechst 33258 revealed the typical apoptotic condensation and fragmentation of chromatin. Al was also found to be neurotoxic, causing first (4–6 days) abnormal clustering and aggregation, and later (8–12 days) neuronal death. Interestingly, Al neurotoxicity occurred in neuroglial cultures containing approximately 10% astrocytes but not in near-pure neuronal cultures containing only 1% astrocytes. Staining of co-cultured cells with Hoechst 33258 showed apoptotic condensation and fragmentation of chromatin in Al-treated astrocytes but not in co-cultured neurons. This study demonstrates that Al can induce the apoptotic degeneration of astrocytes, and that this toxicity is critical in determining neuronal degeneration and death. Al-mediated apoptosis of cultured astrocytes may be also a valuable model system to study the mechanisms underlying apoptosis in glial cells [47].

## 4.14 Protection of Oligodendrocytes by Heat Shock Proteins (HSP) or Stress Proteins from Oxidative Stress

Heat shock proteins (HSP) or stress proteins serve as biomarkers to identify the contribution of stress situations underlying the pathogenesis of degenerative diseases of the CNS. Goldbaum et al. have analyzed by an immunoblot technique the constitutive and inducible occurrence of stress proteins in cultured rat brain oligodendrocytes subjected to heat shock or oxidative stress exerted by hydrogen peroxide, or a combination of both [48]. The data demonstrate that oligodendrocytes constitutively express HSP32, HSP60, and the cognate form of the HSP70 family of proteins, HSC70. After heat shock, HSP25, alpha B-crystallin, and HSP70 were upregulated, while after oxidative stress the specific induction of HSP32 and alpha B-crystallin was observed. HSP32 represents heme oxygenase 1 (HO-1), a small stress protein with enzymatic activity involved in the oxidative degradation of heme which participates in iron metabolism [49, 50]. The presence of the iron chelators phenanthroline or desferroxiamine (DFO), which previously has been shown to protect oligodendrocytes from oxidative stress-induced onset of apoptosis, caused a marked stimulation of HSP32 without affecting HSP70. This indicates that DFO possibly exerts its protective role by directly influencing the antioxidant capacity of HO-1. In summary, HSP in oligodendrocytes are differentially stimulated by heat stress and oxidative stress. HO-1 has been linked to inflammatory processes and oxidative stress [51], its specific up-regulation after oxidative stress in oligodendrocytes suggests that it is an ideal candidate to investigate the involvement of oxidative stress in demyelinating diseases.

## 5 Conclusion

Mammalian cells take up iron (Fe) via three kinds of iron uptake systems: transferrin (Tf) receptor-mediated endocytosis, redox, and Tf-independent Fe uptake systems. As Al has similar chemical properties to Fe but is a simple trivalent cation incapable of redox changes, the cells can take up Al by Tf receptor-mediated endocytosis and Tf-independent iron uptake systems. Mammalian cells modulate the intracellular levels of various metals by interaction between iron regulatory proteins (IRP-1 and -2) and Fe-responsive elements in untranslated regions of the mRNAs of Fe metabolism related proteins. Al also directly binds a cellular iron sensor, IRP-1, to alter the function of its aconitase and mRNA binding proteins in the same way as Fe does. Al also antagonizes the Fe-induced decrease in IRE binding activity of IRP-2 and inhibits Fe-induced oxidation of IRP-2 *in vitro*. Therefore, Al stabilizes IRP-2 by interfering with the iron-catalyzed oxidation, which results in perturbation of iron metabolism. Al may involve selective targeting and removal of oxidatively modified proteins that may contribute to the turnover of many proteins degraded by the proteasome [27]. The IRP-1 is present at similar levels in both AD and control brains. In contrast, IRP-2 co-localizes with redox-active Fe in NFT, senile plaque neurites, and neuropil threads. Hence, alteration in IRP-2 might be directly linked to impaired iron homeostasis in AD [28]. These new effects of Al in brain cells may be concerned with a cause of Alzheimer's disease. Apoptosis pathways in brain cells are induced by Al and/or Fe to result in neuronal and/or glial cell death. On the other hand, heat shock and stress proteins are induced for cell protection. The apoptosis process by oxidative stress is inhibited by Fe chelator or suppressed by some of stress proteins (e.g., heme oxygenase-1) concerned with Fe metabolism. Therefore, it is considered to be important to clarify at, the level of molecular biology, the relations between apoptosis process, oxidative stress, and Fe metabolism in neuronal and glial cells. The more information we have on Fe metabolism, the more will we be able to deeply understand a new effect of Al in mammalian cells.

**Acknowledgement.** I am grateful to Dr. Tadashi Oshiro for editing of this manuscript.

## 6 References

1. Basset P, Quesneau Y, Zwiller J (1986) Iron-induced L1210 cell growth: Evidence of a transferrin-independent iron transport. Cancer Res 46: 1644–1647
2. Inman RS, Wessling-Resnick M (1993) Characterization of transferrin-independent iron transport in K562 cells. J Biol Chem 268: 8521–8528
3. Kaplan J, Jordan I, Sturrock A (1991) Regulation of the transferrin-independent iron transport system in cultured cells. J Biol Chem 266: 2997–3004
4. Thorstensen K, Romslo I (1988) Uptake of iron from transferrin by isolated rat hepatocyte: A redox-mediated plasma membrane process. J Biol Chem 263: 8844–8850

5. Oshiro S, Nakajima H, Markello T, Krasnewich D, Bernardini I, Gahl WA (1993) Redox, transferrin-independent, and receptor-mediated endocytosis iron uptake system in cultured human fibroblasts. J Biol Chem 268: 21586–21591
6. Randell EW, Parkes JG, Olivieri NF, Templeton DM (1994) Uptake of non-transferrin-bound iron by both reductive and non-reductive processes is modulated by intracellular iron. J Biol Chem 269: 16046–16053
7. Sturrock A, Alexander J, Lamb J, Craven CM, Kaplan J (1990) Characterization of a transferrin-independent uptake system for iron in HeLa cells. J Biol Chem 265: 3139–3145
8. Benjamin L (1994) Genes V. Oxford University Press, New York
9. Gunshin H, Mackenzie B, Berger UV, Gunshin Y, Romero MF, Boron WF, Nussberger S, Gollan JL, Hediger A (1997) Cloning and characterization of a mammalian proton-coupled metal-ion transporter. Nature 388: 482–488
10. Fleming MD, Romano MA, Su MA, Garrick LM, Garrick MD, Andrew NC (1998) Nramp2 is mutated in the anemic Belgrade (b) rat: Evidence of a role for Nramp2 in endosomal iron transport. Proc Natl Acad Sci USA 95: 1148–1153
11. Ganrot PG (1986) Metabolism and possible health effects of aluminum. Environment Health Perspect 65: 363–441
12. Roskams AJ, Connor JR (1990) Aluminum access to the brain: a role for transferrin and its receptor. Proc Natl Acad Sci USA 87: 9024–9027
13. Oshiro S, Nakamura Y, Ishige R, Hori M, Nakajima H, Gahl, WA (1994) Reduction site of transferrin-dependent and transferrin-independent iron in cultured human fibroblasts. J Biochem 115: 849–852
14. Oshiro S, Kawahara M, Shirao M, Muramoto K, Kobayashi K, Ishige R, Nozawa K, Hori M, Cai Y, Kitajima S, Kuroda Y (1998) Aluminum taken up by transferrin-independent iron uptake affects the iron metabolism in rat cortical cells. J Biochem 123: 42–46
15. Oshiro S, Kawahara M, Kuroda Y, Zhang C, Cai Y, Kitajima S, Shirao M (2000) Glial cells contribute more to iron and aluminum accumulation but are more resistant to oxidative stress than neuronal cells. Biochim Biophys Acta 1502: 405–414
16. Takeda A, Devenyi A, Connor JR (1998) Evidence for non-transferrin-mediated uptake and release of iron and manganese in glial cell cultures from hypotransferrinemic mice. J Neurosci Res 51: 454–462
17. Oshiro S, Nozawa K, Cai Y, Hori M, Kitajima S (1998) Characterization of a transferrin-independent iron uptake system in rat primary cultured cortical cells. J Med Dent Sci 45: 171–176
18. Abreo K, Abreo F, Sella ML, Jain S (1999) Aluminum enhances iron uptake and expression of neurofibrillary tangle protein in neuroblastoma cells. J Neurochem 72: 2059–2064
19. Klausner RD, Rouaoult TA, Harford JB (1993) Regulating the fate of mRNA: the control of cellular iron metabolism. Cell 72: 19–28
20. Theil EC (1994) Iron regulatory elements (IREs): a family of mRNA non-coding sequences. Biochem J 304: 1–11
21. Henze MW, Kuhn LC (1996) Molecular control of vertebrate iron metabolism: mRNA-based regulatory circuits operated by iron, nitric oxide, and oxidative stress. Proc Natl Acad Sci USA 93: 8175–8182
22. Hail DJ, Rouaoult TA, Harford JB, Kennedy MC, Bondin GA, Beinert H, Klausner RD (1992) Cellular regulation of the iron-responsive element binding protein: disassembly of the cubane iron-sulfur cluster results in high-affinity RNA binding. Proc Natl Acad Sci USA 89: 11735–11739
23. Hail DJ, Rouaoult TA, Tang CK, Chin J, Harford JB, Klausner RD (1992) Reciprocal control of RNA-binding and aconitase activity in the regulation of the iron-responsive element binding protein: role of the iron-sulfur cluster. Proc Natl Acad Sci USA 89: 7536–7540

24. Oshiro S, Nozawa K, Hori M, Chun Zhang, Hashimoto Y, Kitajima S, Kawamura K (2002) Modulation of iron regulatory protein-1 by various metals. Biochem Biophys Res Comm 290: 213–218
25. Guo B, Yu Y, Leibold EA (1994) Iron regulates cytoplasmic levels of a novel iron-responsive element-binding protein without aconitase activity. J Biol Chem 269: 24252–24260
26. Iwai K, Drake SK, Wehr NB, Weissman AM, LaVaute T, Minato N, Klausner RD, Levine RL, Rouault TA (1998) Iron-dependent oxidation, ubiquitination, and degradation of iron regulatory protein 2: implications for degradation of oxidized proteins. Proc Natl Acad Sci USA 95: 4924–4928
27. Yamanaka K, Minato N, Iwai K (1999) Stabilization of iron regulatory protein 2, IRP2, by aluminum. FEBS Lett 462: 216–220
28. Smith MA, Wehr K, Harris PLR, Siedlak SL, Connor JR, Perry G (1998) Abnormal localization of iron regulatory protein (IRP) in Alzheimer's disease. Brain Res 788: 232–236
29. Abreo K, Abreo F, Sella M, Jain S (1999) Aluminum enhances iron uptake and expression of neurofibirillary tangle protein in neuroblastoma cells. J Neurochem 72: 2059–2064
30. Shin R-W, Lee VM-Y, Trojanowsky JQ (1994) Aluminum modifies the properties of AD PHFtau proteins *in vivo* and *in vitro*. J Neurosci 14: 7221–7233
31. Fleming J, Joshi JG (1987) Ferritin: isolation of aluminum-ferritin complex from brain. Proc Natl Acad Sci USA 84: 7866–7870
32. Joshi JG, Fleming JT, Dhar M, Chauthaiwale VJ (1995) A novel ferritin heavy chain messenger ribonucleic acid in the human brain. J Neurol Sci 134: 52–56
33. Clauberg M, Joshi JG (1993) Regulation of serine protease activity by aluminum: implications for Alzheimer disease. Proc Natl Acad Sci USA 90: 1009–1012
34. Mantyh PW, Ghilardi JR, Rogers S, DeMaster E, Allen CJ, Stimson ER, Maggio JE (1993) Aluminum, iron, and zinc ions promote aggregation of physiological concentrations of beta-amyloid peptide. J Neurochem 61: 1171–1174
35. Kawahara M, Muramoto K, Kobayashi K, Mori H, Kuroda Y (1994) Aluminum promotes the aggregation of Alzheimer's amyloid beta-protein *in vitro*. Biochem Biophys Res Commun 198: 531–535
36. Good PF, Perl DP, Bierer LM, Schmeidler J (1992) Selective accumulation of aluminum and iron in the neurofibrillary tangles of Alzheimer's disease: a laser microprobe (LAMMA) study. Ann Neurol 31: 286–292
37. Mashiko S, Suzuki N, Koga S, Nakano M, Goto T, Ashino T, Mizumoto I, Inaba H (1991) Measurement of rate constants for quenching singlet oxygen with a Cypridina luciferin analog (2-methyl-6-[*p*-methoxyphenyl]-3,7-dihydroimidazo [1,2-*a*]pyrazin-3-one) and sodium azide. J Biolumin Chemilumin 6: 69–72
38. Koga S, Nakano M, Uehara K (1991) Mechanism for the generation of superoxide anion and singlet oxygen during heme compound-catalyzed linoleic acid hydroperoxide decomposition. J Biolumin Chemilumin 289: 223–229
39. Siesjo BK (1984) Cerebral circulation and metabolism. J Neurosurg 60: 883–908
40. Makar TK, Nedergaad M, Preuss A, Gelbard AS, Perumal Cooper AJL (1994) Vitamin E, ascorbate, glutathione, glutathione disulfide, and enzymes of glutathione metabolism in cultures of chick astrocytes and neurons: Evidence that astrocytes play an important role in antioxidative processes in the brain. J Neurochem 62: 45–53
41. Campbell A, Bondy SC (2000) Aluminum induced oxidative events and its relation to inflammation: a role for the metal in Alzheimer's disease. Cell Mol Biol 46: 721–730
42. Bondy SC, Kirstein S (1996) The promotion of iron-induced generation of reactive oxygen species in nerve tissue by aluminum. Mol Chem Neuropathol 27: 185–194
43. Lukiw WJ, Bazan NG (1998) Strong nuclear factor-kappaB-DNA binding parallels cyclooxygenase-2 gene transcription in aging and in sporadic Alzheimer's disease superior temporal lobe neocortex. J Neurosci Res 153: 583–592

44. Savory J, Rao JK, Huang Y, Letada PR, Herman MM (1999) Age-related hippocampal changes in Bcl-2: Bax ratio, oxidative stress, redox- active iron and apoptosis associated with aluminum-induced neurodegeneration: increased susceptibility with aging. Neurotoxicology 20: 805–817
45. Guo GW, Liang YX (2001) Aluminum-induced apoptosis in cultured astrocytes and its effect on calcium homeostasis. Brain Res 888: 221–226
46. Ghribi O, DeWitt DA, Forbes MS, Herman MM, Savory J (2001) Co-involvement of mitochondria and endoplasmic reticulum in regulation of apoptosis: changes in cytochrome c, Bcl-2 and Bax in the hippocampus of aluminum-treated rabbits. Brain Res 903: 66–73
47. Suarez-Fernandez MB, Soldado AB, Sanz-Medel A, Vega JA, Novelli A, Fernandez-Sanchez MT (1999) Aluminum-induced degeneration of astrocytes occurs via apoptosis and results in neuronal death. Brain Res 24: 835, 125–136
48. Goldbaum O, Richter-Landsberg C (2001) Stress proteins in oligodendrocytes: differential effects of heat shock and oxidative stress. J Neurochem 78: 1233–1234
49. Poss KD, Tonegawa S (1997) Reduced stress defense in heme oxygenase 1-deficient cells. Proc Natl Acad Sci USA 94: 10925–10930
50. Poss KD, Tonegawa S (1997) Heme oxygenase 1 is required for mammalian iron reutilization. Proc Natl Acad Sci USA 94: 10919–10924
51. Oshiro S, Takeuchi H, Matsumoto M, Kurata S (1999) Transcriptional activation of heme oxygenase-1 gene in mouse spleen, liver and kidney cells after treatment with lipopolysaccharide or hemoglobin. Cell Biol Int 23: 465–474

# The Complexity of Aluminum-DNA Interactions: Relevance to Alzheimer's and Other Neurological Diseases

Suram Anitha, Kosagi Sharaf Jagannatha Rao

Department of Biochemistry and Nutrition, Central Food Technological Research Institute, Mysore – 570 013, India
*E-mail: kjr4n@yahoo.com*

The DNA molecule is dynamic and polymorphic in nature. Although the right-handed B-form is the most predominant conformation, non-B-DNA conformations also occur in biological systems, for example, Z-DNA, triple helix, tetraplexes, hairpin, cruciform etc., under various physiological conditions. Extensive studies on handedness and various high-ordered structures were carried out for the oligonucleotides, but little is known about genomic DNA topology with respect to non-Watson-Crick right-handed B-DNA forms, their functional ability and possible implication in pathogenic features in the brain, if any?. Alzheimer's disease (AD) is a complex neurological disorder in which etiological factors are not clearly known, although unproven hypotheses have included head trauma, aluminum toxicity and infectious agents. Studies have shown that there is DNA damage, chromatin condensation, and altered gene expression in AD brain. We, for the first time, showed evidence for the presence of a B-Z intermediate form and Z-DNA conformation in hippocampus region in moderately affected and severely affected AD, respectively. Our group also reported for the first time the role of aluminum in modulating DNA topology in d(GCCCATGGGC), d(CCGGGCCCGG), CCG-repeats, supercoiled DNA and calf thymus genomic DNA. We propose a new hypothesis on the role of Al in neurodegeneration of Alzheimer's disease. The relevance of altered DNA topology in Alzheimer's brain, with reference to altered gene expression, nucleosome organization and assembly, DNA susceptibility to damage and terminal differentiation will be discussed.

**Keywords:** Aluminum, DNA, Helical transitions, Spectroscopy, Alzheimer's disease

## 1 Introduction

Metal ions play an important role in the stabilization of DNA and in the catalytic activity of DNA polymerases [1]. Transition metals like Mn and Co, etc., at higher ionic strength are found to be mutagenic and carcinogenic in nature and also change the helical nature of the DNA [1–6]. Aluminum (Al) is reported to be associated with neurological disorders and to cause cellular toxicity through its interaction with DNA in cells [7–9].

Aluminum plays a role in causing the genetic instability associated with the repeats by modulating the conformation of the repeats. Our studies for the first time showed that the Al concentration in serum of fragile-X affected patients was elevated and provided a clue on its involvement in causing genomic instability [8, 9]. Aluminum (Al) is one of the strongly suspected factors in neurodegenerative disorders like Alzheimer's, Parkinsonism and Guam-Parkinsonism/amyotrophic lateral sclerosis [10–13]. Studies from our laboratory [14, 15] and elsewhere [16] showed that Al preferentially binds to DNA. However, our studies showed that Al strongly binds to AT*-rich sequences with no conformational transition [15], while it induces the helical transition in GC*-rich oligomers from the Z-A DNA conformation [14]. Our main question is: does the topology of CCG repeats play a vital role fragile-X neurodisorders through the modulation of transcription of FMR1 gene? We hypothesize that DNA topology in these repeats has an important role in FMR1 gene expression. Keeping the above complexity in view, the topology of $(CCG)_{12}$-long repeats and the role of Al in modulating CCG topology was studied. Further, to hypothesize the relevance of these topological changes the transcription of the FMR1 gene was investigated. Hitherto, there are no studies in this direction.

## 2 Aluminum Speciation Chemistry and Its Relevance to DNA Binding

The speciation chemistry of Al is especially important in the experimental design of research investigations into Al toxicity. Al speciation in the stock solutions must be evaluated, since it hydrolyzes readily and at pH 7.0 there is a strong tendency for the precipitation of $Al(OH)_3$ which makes the preparation of Al stock solutions difficult. Aluminum is a strong hydrolyzing element, and is generally insoluble at neutral pH. Its solubility is enhanced under acidic or alkaline conditions, and in the presence of complexing agents. In aqueous solution at pH < 5.0, aluminum exists as an octahedral hexahydrate, $[Al(H_2O)_6]^{3+}$, usually abbreviated as $Al^{3+}$. As the solution becomes less acidic, $[Al(H_2O)_6]^{3+}$ undergoes successive deprotonations to yield different species such as $[Al(OH)]^{2+}$, $[Al(OH)_2]^+$ and $Al(OH)_3$. Neutral solutions give an $Al(OH)_3$ precipitate that re-dissolves, owing to the formation of tetrahedral aluminates, $[Al(OH)_4]^-$, the primary soluble Al(III) species at pH > 6.2, the biological pH. The overall deprotonation reactions will take place according to the following equation.

$$[Al(H_2O)_6]^{3+}[Al(H_2O)_5(OH)]^{2+}[Al(H_2O)_4(OH)_2]^{+}Al(OH)_3[Al(OH)_4]^{-} \quad (a)$$

One cannot compute the soluble Al concentration of the solution simply by adding a known quantity of an Al compound to water, without taking hydrolysis reactions into account. For example, when inorganic Al salts such as chloride, sulfate, hydroxide or perchlorite are dissolved in water at a calculated concentration of 10 mM, the exact Al concentration after pH adjustment and filtration is about 50 μM. The use of aluminum lactate or aluminum aspartate, however, increases the soluble Al concentration to ~55–330 μM, and use of aluminum maltolate or gluconate increases the soluble Al concentration to 4000–6000 μM. Aluminum maltolate is particularly suitable for toxicological studies because of its defined molecular structure in solution, its neutral charge, high solubility and hydrolytic stability at pH 7.0. For in vitro studies to investigate the Al interaction with DNA, it is of utmost importance to have a higher free $Al^{3+}$ ion concentration at physiological pH. This cannot be achieved by using any inorganic Al salt due to the insolubility problem at physiological pH. Aluminum maltolate provides higher free Al concentrations over a wide range of pH, rendering it the most efficient compound for biological studies. At this pH, the total soluble Al ion, 'r' is given by $(Al^{3+})/[(Al^{3+} + Al(OH)^{2+} + Al(OH)_2^+ + Al(OH)_4^-]$ and $1/r = 1 + [(10^{5.5}/H^+) + (10^{-11.1}/(H^+)^2 + (10^{-23.2}/(H^+)^4]$. Using the above distribution pattern of Al species at pH 7.4, free Al concentration is expressed [8] according to the following equation.

$$Al(OH)_3 + 3\,H^+ \rightarrow Al^{3+} + 3\,H_2O \text{ and } K.sol. = (Al^{3+})/(H^+) = 10^{-7} \text{ to } 10^{-9} \quad (b)$$

# 3
# Alzheimer's Disease (AD) Brain DNA

The DNA molecule is dynamic and polymorphic in nature and it can form various structures like right-handed, left-handed, tetraplexes, cruciform and hairpin [17]. Information on the conformational manifestation of eukaryotic DNA or their biological significance with reference to gene regulation and expression is limited. This is of considerable interest as biological function is often correlated with structure at molecular levels. The changes in the genetic material of Alzheimer's disease, a devastating neurodegenerative disorder, so far documented are: mutations in genes, amyloid precursor protein (APP), presenilin-1 (PS-1) and presenilin-2 (PS-2) [18], shift in the DNA from a potentially transcribable euchromatic state to a condensed and probably heterochromatic state [19], oxidative damage as revealed by damaged products, namely 8-OH-guanosine and 8-OH-cytosine [20–22] and genetic imbalance caused by the tetraploid status of the genome by the way of unusual replication before the cell death [23]. We for the first time demonstrated the presence of the left-handed Z-DNA conformation in severely affected AD (SAD) brain, whereas control young, and aged brain DNA have the usual B-DNA conformation. Moderately affected AD brain DNA has modified B-DNA conformation with a small positive shoulder peak at 295 nm as revealed by circular Dichroism spectroscopy (Fig. 1). The binding of ethidium bromide is high for moderately affected AD DNA, low for severely affected AD DNA compared with control young and control aged DNA. Melting temperature (Tm) data indicated that normal young and normal aged brain DNA have Tm

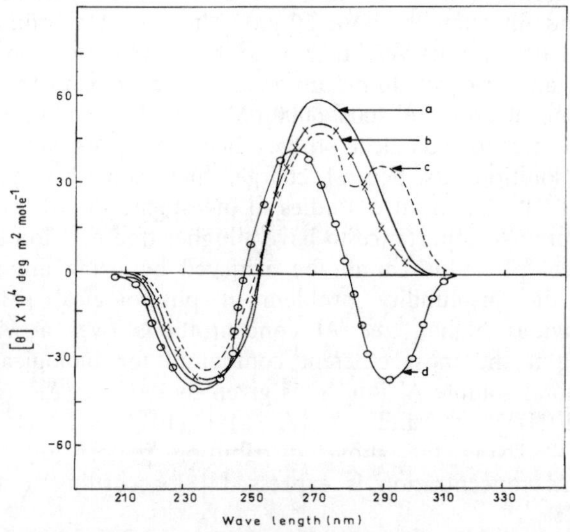


**Fig. 1.** Circular dichroism of genomic DNA in 0.01 M HEPES (pH 7.4). *a*: Normal young DNA, *b*: normal aged DNA, *c*: moderately affected AD DNA, and *d*: severely affected AD

of 65 °C and 59 °C, respectively. SAD-DNA, as expected for Z-DNA, has a high Tm (79 °C) since Z-DNA is a stable and high-energy conformer of B-DNA. MAD-DNA has an unusual biphasic melting profile, which is biologically a rare phenomenon with two Tm values, 54 °C and 79 °C, and thus MAD-DNA is showing a clear-cut violation of the all-or-none response of long duplex DNA. These findings could explain for the first time unusual phenomenon like nucleosome disassembly, G*-specific oxidation and altered gene expression.

## 4 Biological Implications of Z-DNA and Its Neurochemical Implications in Relevance to Alzheimer's Disease

The significant pathophysiological events hypothesized in Alzheimer's pathology are, distorted nucleosome formation, terminal differentiation and DNA damage. There are no explanations at the cellular level as to why AD brain cells adopt these unusual phenomenon. Our novel finding on the presence of Z-DNA in Alzheimer's brain could first be able to clarify the reasoning for the above phenomenon. Brief structural information on Z-DNA is as follows: Z-DNA is a condensed DNA with left-handed conformation. Z-DNA has only the minor groove and thus altered ligand binding ability. Z-DNA is basically classified as ZI and ZII types. ZI and ZII are classified based on the phosphate group orientation. In the ZI conformation, two hydrogen atoms of the guanine N2 amino group are involved in hydrogen bond formation: one bond to cytosine O2 to which it is paired and the other to a water molecule. This water molecule, in turn, forms a hydrogen bond to a phosphate oxygen linked to the deoxyguanosine 3-hydroxy group. Thus, ZI conformations will have this bridging water molecule. In the ZII conformation there is also a hydrogen-bonding bridge. But this involves a chain of two water molecules linking the N2 amino group and the phosphate oxygen. Furthermore, the distances between the two closest phosphate groups in ZI and ZII were 7.7 and 8.6 Å, respectively. The P · P separation across the double helix is 17.4 Å, for B-DNA. The P · P separation values for ZI and ZII were 14.5 and 16.2 Å, respectively [17]. This provided a clue that the ZII conformation may be an interface of B- and Z-DNA. This provides a clue that the DNA present in moderately affected AD brain may be in the ZII conformation and this may be the reason for the biphasic the Tm more EtBr binding and a strong shoulder positive peak at 290 nm.

## 5 Z-DNA and Nucleosome Formation

A great deal of information is available on the role of point mutations, abnormal proteins (A$\beta$, presenilin, phospharylated aggregated tau) and various environmental factors [aluminum (Al), head injury, etc.] involved in AD neurodegeneration [24]. Limited headway was made in the therapeutic aspects of AD because of the complexity of the disorder. There are few studies concerning modulations in the genetic material. Studies have shown DNA fragmentation through oxidative damage [21], point mutations and changes in the integrity of chromatin structure [18] and no information on the helical nature of DNA. Crapper et al. [18] made an interesting finding with reference to chromatin structure. They reported a shift in the folding of DNA from a potentially transcribable euchromatic state to a condensed and probably heterochromatic state. This change may play a role in altering the chromatin conformation, and result in a reduced availability of genetic information to endogenous RNA polymerases. They have also indicated a reduced accessibility of micrococcal nuclease to the nucleosome linker region in Alzheimer brain chromatin. Lukiw et al. [25] showed that chromosomal linker histones are also considered to be partly responsible for the packing of chromatin into inactive, condensed states and thereby act as general gene repressors. They also observed an inverse relation between the abundance of linker histones and gene activity in Alzheimer's disease. They indicated the possible nucleosome exclusion but did not speculate about the reason for the exclusion. It was reported that Z-DNA formation excludes nucleosome formation [26]. Furthermore, Z-DNA conformation has implications in nucleosome organization and assembly, thus affecting the functional ability of chromatin [26]. DNA bending ability is a crucial factor in nucleosome assembly. In B-DNA, negatively charged groups are evenly distributed, thus favoring positive histone interaction and, in turn, good nucleosome assembly. In Z-DNA, negative charges are not uniform due to the zigzag array of backbone phosphates. Hence, the poly(G-C) or poly(G-Cmet) in the Z-DNA conformation excludes nucleosome formation. This charge distribution is not favorable for stable nucleosome structures. There is only one report stating that nucleosomes can also be prepared by cores containing the Z-form of poly(dG)-(5meC). It was shown by Garner and Felsenfled [26] that histone wrapping around left-handed Z-DNA is energetically costly, hence it is not possible to get a stable nucleosome in the cell system. The extent of DNA adopting the Z-conformation *in vivo* is an unsettled issue, since most of DNA is organized in nucleosomes. It is certain that only relatively short stretches of DNA will exist in the Z-form in the cell. These short Z-DNA conformation stretches are sufficient to bring about an altered pattern of nucleosome placement [26]. All these reports showed that most non-B-DNA conformations are likely to have both an altered twist and reduced ability to enter the nucleosomes. This biophysical explanation will provide molecular hypothetical evidence that Z-DNA conformations in AD cells have great relevance in nucleosomes' placement and gene's functional ability. This argument is likely to open up new innovations in understanding the role of DNA topology in neurodegeneration in AD.

## 6 Z-DNA and Terminal Differentiation

These studies provided a clue that AD brain cells possibly undergo terminal differentiation pattern and likely to prefer to the Z-DNA conformation in their genomic DNA. Z-DNA is considered to be involved in both transcriptional activation and inactivation depending on the cell's physiological condition [27]. Recently, Gagna et al. [28] reported that left-handed Z-DNA is speculatively a transcriptional enhancer and might be directly influenced by the destructive effects of terminal differentiation, thus making DNA transcriptionally inactive. The nicking-denaturation of double-stranded Z-DNA could possibly initiate and enhance terminal differentiation [28]. It is important to see the transcriptional ability of the Z-DNA conformation in AD cells undergoing neurodegeneration. This information will provide greater insight in understanding the molecular mechanism involved in AD pathogenesis.

## 7
## Z-DNA and DNA Damage

DNA damage plays an important role in the genomic instability and neuronal cell death. Studies on DNA damage in AD brain showed that the levels of oxidatively damaged cytosine and guanine (G*) bases, namely 5-hydroxycytosine and 8-hydroxyguanine were significantly higher in AD brain over normal brain [20]. The main questions to answer are why G*-specific oxidation is more in AD brain and does DNA's helical nature have any role in this specificity? It was theoretically postulated that the G* base present in DNA would be more susceptible to hydroxy radical-induced DNA damage if the conformation of DNA is Z rather than B or A because of greater exposure of *G-C bases [29]. Our present finding on the presence of Z-DNA in AD brain will support the experimental findings on the presence of G-C*-rich oxidized products in AD brain [20].

## 8
## Our Hypothesis

We hypothesize (Fig. 2) that the abnormal proteins expressed in AD, trace elemental homeostatic imbalance, methylation of GC* sequences in genomic DNA and oxidative stress all together might have a role in Z-DNA conversion. Whether an altered DNA structure in the hippocampus region reflects the pathogenic event or represents as an adaptation of terminal differentiation cannot be ascertained at present. But it is quite interesting that the DNA in SAD is showing both the Z-conformation and also the typical apoptotic pattern. An apoptotic DNA fragmentation pattern might be one of the crucial steps in B-Z DNA conversion since the fragmentation reduces the high activation energy barrier required to induce this conformation. Trumbore et al. [30] reported a metal-assisted B-Z conformation in DNA which undergoes strand breaks due to radiation-induced DNA damage. The smaller metal ion concentration could be better induce this conversion in DNA with strand breaks than in the absence of strand breaks. They attributed this to a reduction in the high activation energy barrier in the metal ion-driven B-Z conformation.

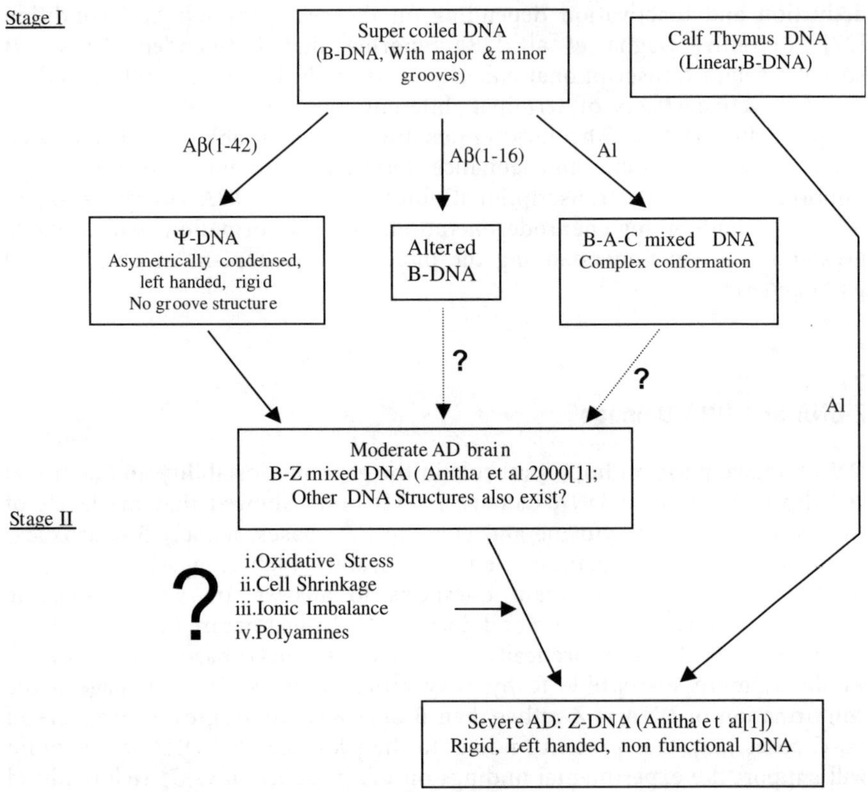


**Fig. 2.** Possible role of Al, A$\beta$(1–42) and A$\beta$(1–16) in modulating. supercoiled DNA topology: relevance to AD

It also provides a clue to why SAD-DNA could go to the Z-conformation from B, probably due to DNA fragmentation as observed in our results. We could not observe the Z-conformation in MAD because there is no DNA fragmentation. Recently, it was shown that lysine-rich proteins favor the B-Z DNA transition and further that methylated DNA can be easily converted from B to Z by the lysine rich proteins [31]. It is interesting to hypothesize that AD-specific proteins like A-beta, phospharylated Tau are rich in lysine. Probably they may have a role in inducing the B-Z transition in AD brain and further work is in progress to validate this hypothesis.

## 9 Aluminum-Induced Topological Changes in DNA

Al binding to DNA has been studied by number of groups. But the binding pattern and mechanism of the Al-induced changes in DNA topology is little understood due to the Al speciation chemistry problem as explained above. In this following explanation, we will highlight the complexicity of Al chemistry and its relevance to DNA structural studies.

a) Al interaction with calf thymus DNA: Karlik et al. [16] studied the interaction of Al with DNA using thermal denaturation, circular dichroism (CD), and fluorescent dye binding techniques as a function of pH and temperature. They reported three distinguishable complexes when DNA is denatured at pH 5.0–7.5 and at Al to DNA molar ratios of 0.07. At neutral pH complex I appears and stabilizes DNA. The complex II was observed at acidic pH and it stabilizes a fraction of the DNA double-helical molecule and thus produces intrastrand cross-linking. It is interesting to note from their findings that complex III occurs at all pHs and that this complex is characterized by a non-cooperative melting profile and cross-linking at pH below 6.0. The complexes II and III can be denatured by treatment with either EDTA or high NaCl. The properties of complexes I and II are consistent for a DNA complex with $Al(OH)^{2+}$ and $Al^3$, respectively. However, complex III depends on the ratio of $Al^{3+}/(OH)^{2+}$ in solution. Al-DNA complexes differ from other metal-DNA complexes in melting profile. It was highlighted that more than one form of DNA can exist at any time in the presence of Al. Ahmed et al. [32] studied the interaction of Al and Ga with calf thymus DNA using FT-IR. They clearly showed that at low metal ion concentrations (r = 1/80, 1/40), both cations bind mainly to the backbone $PO_2$ group and the guanine N7 site of the G-C base pairs as a function of the chelation property. The chelation character of the cation is evidenced from major shift and intensity increase of the phosphate antisymmetric stretch at 1222 $cm^{-1}$ and mainly the guanine band at 1717 $cm^{-1}$. The A-T base pairs were perturbed at higher concentrations of cation indicating major helix destabilization. All these FT-IR studies indicated that a major reduction of the right-handed B-DNA structure occurs in favor of A-DNA formation. Rao and Divakar [33] reported the strong binding of Al to calf thymus DNA with a strong binding constant ($10^3$ $M^{-1}$) and did not induce any helicity change (Fig. 3a). We have used NMR and circular dichroism techniques to examine the DNA binding abilities of inorganic salts of Al. Recently, we showed that Al-maltol, a soluble Al complex at neutral pH, could induce the Z-DNA conformation (Fig. 3b) in calf thymus DNA at 1 mM concentration using different spectroscopic techniques (unpublished results).

b) Aluminum cross-linking: Karlik and Eichhorn's [34] studies indicated that Al cross-links with DNA of all base ratios, including poly(dAdT) and poly(dGdC). Al cross-links prevent renaturation in poly(dAdT). Saturation of cross-linking occurs at a 0.4 ratio of Al to nucleotide phosphate, indicating that potentially 80% of the base pairs were Al-bound. However, cross-links were broken at elevated pH and by EDTA treatment.

c) Aluminum binding to chromatin: Walker et al. [35] made interesting observations on the binding of aluminum with chromatin isolated from brain. Aluminum could strongly bind to chromatin and also altered the chromatin

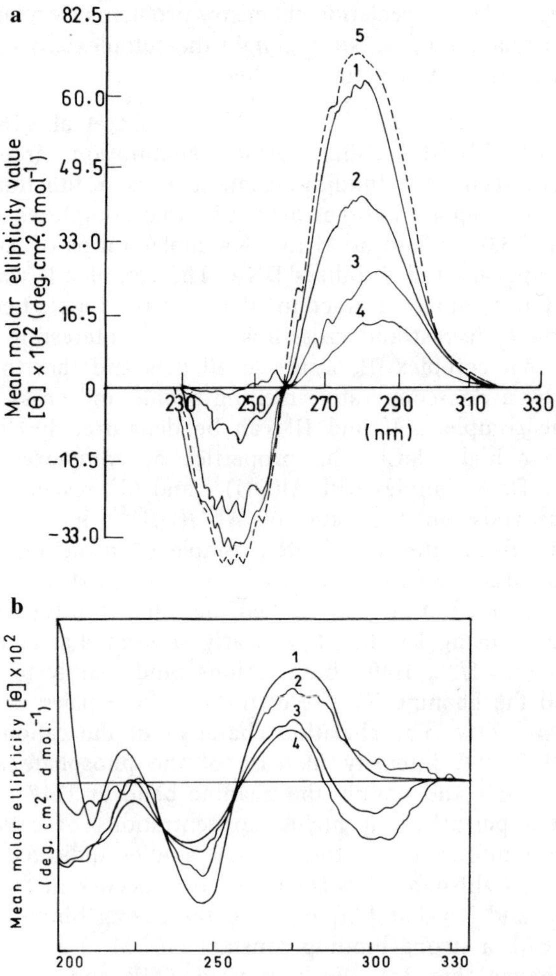


**Fig. 3.** a The binding of aluminum nitrate to calf thymus DNA with no helical transitions. b Role of aluminum maltolate in inducing helicity changes in calf thymus DNA

structure as evidenced by nuclease accessibility. It was also reported by this group that brain chromatin is more sensitive for Al interaction over chromatin isolated from liver.

d) Aluminum complex formation with nucleotides: Al significantly altered the two peak pK values (-$NH_3$, $PO_3$) of ATP, ADP, AMP′-3 and AMP′-5. Proton NMR studies indicated that Al preferentially binds to the H8 site of the adenine series and a molecular structure featuring Al binding to adenine was computed (Fig. 4).

e) Aluminum induced structural changes in sequence-specific oligonucleotides: The CD spectra of d(GCCCATGGGC) and d(GCGTACGC) indicated the B-form of DNA, with a negative band at 240 nm and a major

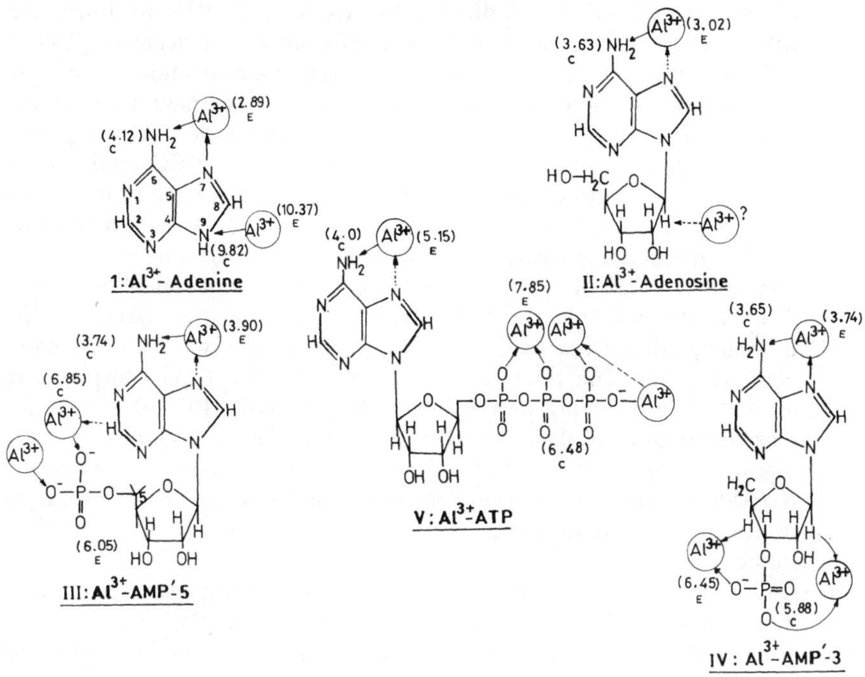


**Fig. 4.** Aluminum binding to adenine and its analogues

positive band at 265 nm. The CD spectra of d(CCGGGCCCGG) indicated the Z-form of DNA with two characteristic negative peaks at 245 and 290 nm, respectively, and a positive peak at 260 nm. The CD spectra of $(CCG)_{12}$ also indicated B-DNA conformation. Al showed strong binding to d(GCGTACGC) with no transition, it converted the Z- to the A-form in d(CCGGGCCCGG) and caused a modified B-DNA conformation in d(GCCCATGGGC). In CCG-repeats on addition of lower concentrations ($10^{-7}-10^{-6}$ M) of Al, there was only a decrease in the magnitude of the positive and negative bands at 275 and 245 nm, respectively. But at higher Al concentration ($10^{-5}$ and $10^{-4}$ M), CCG attained the Z-conformation.

f) Role of Al in modulating triplet repeats structural dynamics: The expansion of DNA-triplet repeats define a new type of mutation in the human genome, namely "dynamic mutation" and they cause genomic instability. The mutation studies have revealed that abnormal expansion of CCG, CGG, CTG, CAG etc., were implicated in more than 12 neurological disorders [36]. GC*-rich triplet repeats CCG, CGG, GCC, GGC are predominantly involved in X-fragile neurodisorder. These repeats are localized both in the coding and non-coding regions of genes [36]. Furthermore, the expansion of these repeats differs depending on where in the gene the repeat tract is located and range from reduction of transcription initiation to protein toxicity.

i) CD spectra of $(CCG)_{12}$ indicated the classical B-DNA conformation with a characteristic positive peak at 275 nm and a negative peak at 245 nm (Fig. 5a). On addition of lower concentrations ($10^{-8}$ to $10^{-6}$ M) of Al, there was only a decrease in the magnitude of the positive and negative peaks at 275 and 245 nm, respectively (Fig. 5b). But at higher Al concentration ($10^{-5}$ and $10^{-4}$ M), CCG attained the Z-DNA conformation with the characteristic two negative peaks at 245 and 290 nm and a positive peak at 260 nm (Fig. 5c and d). The effect of the Al-chelating drug, desferoximine ($10^{-5}$ M), on the Al-CCG complex ($10^{-5}$ M of Al) indicated that the total removal of Al by the drug could not reverse the spectral characteristics back to the B-conformation (free CCG) from the Z-conformation (CCG-Al complex) (Fig. 5e). The total removal of Al from the CCG complex was estimated by dialyzing desferoximine:Al-CCG ($10^{-5}$:$10^{-5}$:$10^{-5}$ M) complex overnight in 0.001 M HEPES buffer. The dialyzed CCG was digested in nitric acid and the Al content in the CCG was estimated by ICPAES, in addition we found no signal in the digested CCG solution indicating total removal of Al from the CCG-oligomer by desferoximine.

ii) C-C Mismatches in CCG repeats: C-C base pairing was detected by monitoring CD changes at 280 nm with reference to pH 5.0 to 7.0 at 0.1 and 0.5 M $Na^+$ concentrations. There was an increase in the magnitude

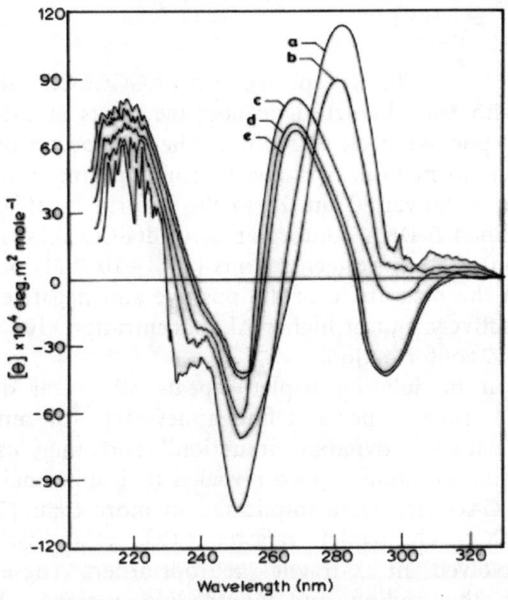


**Fig. 5.** Effect of Al on circular dichroism spectra of $(CCG)_{12}$ $a = (CCG)_{12}$, $b$ = Al ($5\times10^{-7}$ M), $c$ = Al ($5\times10^{-5}$ M), $d$ = Al ($5\times10^{-4}$ M), $e$ = Al ($5\times10^{-4}$ M) + desferoximine ($5\times10^{-4}$ M)

of the long-wave length positive band centered at about 280 nm when the pH was lowered from 7.0 to 5.0 at 0.1 and 0.5 M Na, indicating C-C* mismatch base pair formation. The lower concentration of Al ($10^{-7}$ M) did affect C-C* mismatch base pair formation. But the higher concentration of Al ($10^{-4}$ M) decreased C-C* mismatch base pair formation thus increasing the stability of the DNA conformation.

Our studies for the first time evidenced that Al brought about a helical transition from the B- to the Z-DNA conformation in $(CCG)_{12}$-repeats. The total removal of Al by a chelating drug, namely desferoximine, did not reverse the Z-DNA conformation to B-DNA, indicating the irreversible and stable conformation. The molar concentration of Al in the serum of X-fragile syndrome is $1.8 \times 10^{-4}$ M. In the present investigation, it was observed that $10^{-4}$ M and $10^{-5}$ M of Al could favor the Z-conformation in CCG repeats. It is interesting to mention that the Al concentration recorded ($10^{-4}$ M) in serum could induce the Z-DNA conformation in $(CCG)_{12}$ repeats and thus these findings have physiological significance. Furthermore, the Al-induced Z-DNA conformation is stable as evidenced by the irreversibility of the conformation even after total Al is removed by desferoximine. The Z-DNA conformation in CCG-repeats could act to physically prevent the movement or the binding of RNA polymerase thus modulating transcription. Therefore in fragile X-syndrome, the FMR1 gene might have been turned off due to the formation of Z-DNA due to Al. We developed a hypothesis, which is explained in a later part of the discussion, in strong support of the above concept. Thus, formation of Z-DNA topology in CCG-long repeats has great biological significance in the molecular understanding of the role of the FMR1 gene in X-fragile syndrome and the relevance of Al in neurological disorders.

The long CCG triplet repeats at the FRAXA site are involved in the pathology of fragile X-syndrome [36]. The poor chromatin staining and the large expansions of the CCG-repeat block at the FRAXA site suggested an altered chromatin structure and strong nucleosome exclusion [37]. It is hypothesized that the Z-DNA conformation is also one of the likely factors which favor exclusion of nucleosome formation [38] and this will possibly explain the poor chromatin staining of the fragile site. Z-DNA, being more rigid, cannot bend to favor the formation of nucleosome. Thus, DNA topology is very important for histone protein binding, this binding favors the formation of nucleosome and also is needed for RNA polymerase binding during transcription [39].

## 10
## Our Hypothesis on How FMR1 Gene Regulation Is Modulated by Al-Mediated CCG* Topology

(a) Al-mediated DNA Topological changes from B to Z-DNA: We hypothesize that the molecular involvement of Al in bringing about the B-DNA to Z-DNA conformation may be due to binding of Al to phosphate groups and nitrogenous bases which may affect sugar puckering in dG leading to C2′ to C3′ *endo* pucker and the configuration of the glycosidic bond (torsional angle). Hence, negative super helical tension due to Al can drive local transitions to alternative DNA structures which may affect a particular gene (FMR1). The C-C mismatches induce local flexibility to form non-B-DNA conformation. During creation of the Z-DNA conformation, the intrastrand hydrogen bond distance between C-C mismatches in the B-DNA conformation should decrease due to Al binding (N4-N3, 4.38 Å B-DNA; O2-O2, 3.54 Å Z-DNA) in addition to a change of the torsion angle and sugar puckering in dG. Hence, there is a reduction in C-C mismatches at higher Al concentrations. This may be the reason why desferoximine could not reverse Z- to B-DNA even after removal of Al.

(b) Z-DNA conformational stability pattern: The duplex $(CCG)_{12}$ repeats with C-C* mismatch has a Tm of 48 °C. Al ($10^{-7}$ M) has decreased the Tm to 42 °C. But at higher concentrations of Al ($10^{-4}$ M), CCG has a higher Tm of 52 °C. The reduced Tm at lower concentrations of Al may be due to the transition of intermediary and transition conformations from B-DNA. However, the higher Al concentration caused an increased Tm due to the Z-conformation. The reason might be a transition of nitrogen bases by 180° with respect to the B-conformation and a change of the deoxyribose molecule from a perpendicular to a parallel orientation of the helix axis. During this process, the hydrogen bonds may break due to the more favorable transition energy towards the Z-conformation. Thus, the intermediary duplex needs less energy to melt. Once the Z-conformation is attained, in addition to the usual three hydrogen bonds between guanine and cytosine, a hydrogen bond between N3 cytosine of one chain to $NH_2$-C6-cytosine of the other chain is formed. This is mainly due to the Z-DNA conformation bringing two cytosine molecules (which are in an *anti* conformation) nearer together as compared to the B-conformation and necessitates a higher energy to melt the Z-DNA duplex.

(c) How Al could modulate FMR gene expression: Eukaryotic transcriptional control operates at three levels:

a) modulation of the levels and/or activities of activators and repressors,
b) change in chromatin structure directed by activators and repressors and,
c) direct influence of activators and repressors on assembly of initiation complexes.

We hypothesize that DNA topology is relevant in FMR1 gene regulation (Fig. 6). The promoter-proximal region at the 5′-end of the FMR1 gene [36] contains GC*-rich sequences and also CGG triplet repeats [36]. This gene is hypothesized to be turned off when CCG/CGG repeats are longer than 200 with methylation [39]. In normal conditions, the genomic DNA is in the B-conformation. The CCG repressor protein or p20 binds in the major groove of the cognate site on DNA in a sequence-specific manner and this interferes with the general transcription factors (GTFs) that play a role in the formation of the transcription-initiation complex and regulates gene repression. Thus, the binding and removal of p20 from its cognate site are likely to control FMR1 gene expression. We hypothesize that, under pre-mutation conditions, the GC*-rich promoter region of the FMR1 gene may be in the B-Z transition rather than the usual B-conformation. This B-Z conformation probably elevates p20 protein binding to its cognate site, resulting in the formation of higher levels of FMR1 mRNAs. However, the presence of CGG repeats (100–200) at the 5′-end of the untranslational region of FMR1 mRNA [39] may also be likely to play an important role in gene expression through mRNA production. The possible formation of a duplex between two mRNAs with parallel/anti-parallel and hairpin loop structures at the CGG-rich region likely to interfere with ribosome-mRNA complex formation. Thus, it affects the translation of FMR1 mRNA.

In this report, we for the first time present evidence that Al induces the B- to Z-DNA conformation and that Z-DNA is more energetic compared to B-DNA. In neurodisorder conditions, the trinucleotide repeats will be more than 200 with hypermethylation at the FMR1 operon and this may be in the Z-DNA conformation in the presence of Al. Under such conditions, both p20 and GTFs may not be able to assemble to form the transcription-initiation complex at its cognate site, because there is no major groove in Z-DNA. We also hypothesize that the deprivation of gene expression may be due to the formation of a complex loop structure at the 5′-end of FMR1 mRNA.

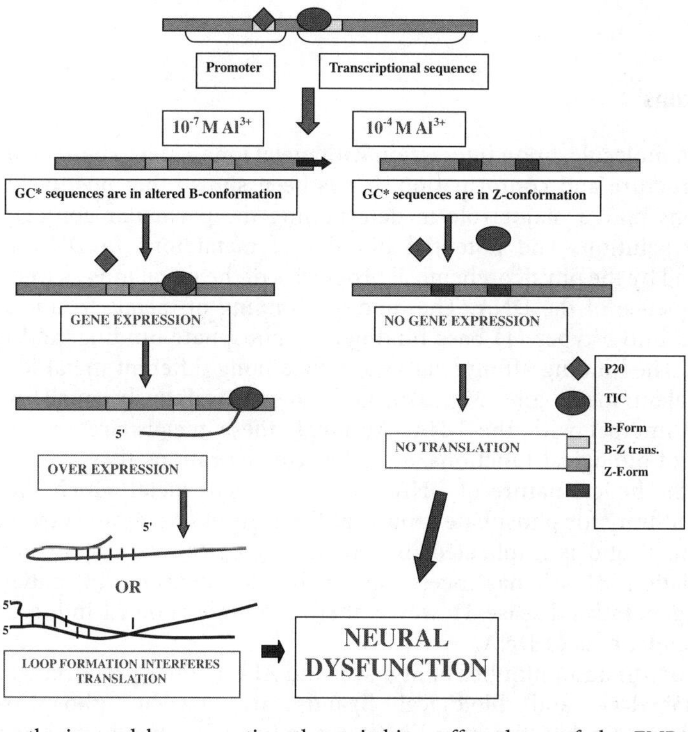


**Fig. 6.** Hypothesis model representing the switching off and on of the FMR1 gene in X-fragile syndrome as a function of DNA structural dynamics

In our model, we concentrate on the direct influence of activators and repressors on the assembly of the transcription-initiation complex to explain the expression of the FMR1 gene. Similar to other eukaryotic genes, these genes do have DNA regions coding for the structural gene along with an upstream promoter-proximal region where transcriptional factors (upstream transcriptional factors or UTFs) bind in a sequence-specific manner and stimulate/inhibit the formation of transcription-initiation complex (general transcription factors or GTFs). The FMR1 gene is known to contain a GC*-rich sequence in the promoter-proximal region (5′-end), specific for UTFs [36]. The suppression/overexpression of the FMR1 gene plays an important role in disease onset. Based on the gene expression during normal and disease conditions, we have postulated that the FMR1 gene is under repression control, since its expression is affected by 5′-(CCG)n-3′ binding protein.

From our studies, it can be assumed that Al is likely to play a role in switching off the FMR1 gene by inducing a helical transition from B- to Z-DNA, in addition to the factors involved in the massive expansion of CCG triplet repeats and hypermethylation. This may provide a clue for the discovery of new drugs which convert Z- to B-DNA. It is hypothesized that chelation of metal ions like aluminum, demethylation of the triplet repeats, and acetylation of histones – if achieved – may make it possible to switch on the FMR 1 gene.

## 11
## Conclusions

The DNA molecule, upon interaction with metal ions, shows significant changes in its structure and conformation. It has been shown that positively charged metal ions play a major role in determining the particular conformation of DNA in solution. The potential binding of metal ions to DNA is greatly influenced by the physico-chemical properties of the metal ions as well as by the base sequence of the DNA. The mode of binding of metals to nucleotides is classified into 3 types (1) base binding, (2) phosphate binding, and (3) sugar binding. The binding affinity varies greatly among different metal ion species. The divalent metal ions Mg, Mn, and Co with relatively small ionic radii strongly interact with the DNA. Although these metals are required for a number of biological functions, at higher concentrations they are reported to change the helical nature of DNA. Al is a trivalent metal which shows great binding affinity for phosphate groups of DNA. Al has no established biological requirement and is implicated in many neurological disorders. Intranuclear accumulation of Al has been shown in the neurons of patients with neurodegenerative disease. However, there is no report on Al-induced changes in the handedness of DNA.

The coordination number of Mg, Mn, and Al is 6, their preferred geometries are octahedral, and biological ligands are oxygen, phosphates, and carboxylates. Co has the coordination number of 4, tetrahedral geometries, and the biological ligands are sulfur and thiolates. The ionic radii (Å) are:

Mn (0.97, 0.81); Co (0.72); Mg (0.86), and Al (0.58). The solvent exchange rates (Kex $S^{-1}$) for inner-sphere water molecules in metal ions are; Al($10^0$); Mg ($10^5$); Co ($10^{5.5}$), and Mn($10^{6.7}$) [21].The order of increasing rate constants in acidic solution is, Al $\ll$ Mg < Co < Mn. Ligand exchange rates take on special importance for Al because they are slow and the system may not be at equilibrium. Mg, Mn, and Co have around $10^5$ faster exchange rate over Al. These differential characteristics of the metals play a crucial role in metal-DNA interactions.

Regarding the stability of DNA altered by aluminum, it is attributed that, in general, metal ions bind with greater affinity as a result of higher charge. The binding of positively charged counterions by nucleic acids is a natural consequence of a polyanionic sugar-phosphate backbone. Metal ions alleviate electrostatic repulsion between phosphates, thereby stabilizing base-pairing and base-stacking. This is most clearly evidenced by the increase in melting temperature and hypochromism, respectively. In contrast, transition metals that have higher affinity for heteroatoms on the bases tend to lower base-pairing and stacking, and so destabilize the double helix and Tm values come down. The alkali or alkali earth metals neutralize repulsive interactions between the negatively charged sugar-phosphate backbones and stabilize the interactions between basepairs. This further stabilizes the hydrogen bonding that pairs the two strands, resulting in higher melting temperatures and leads to enhanced base stacking with increased hypochromism. The potential reactive sites found on the DNA molecule are (a) heterocyclic nitrogen atoms, (b) exocyclic carbonyls on the purine and pyramidine bases, and (c) the phosphate oxygens. The $PO_3^-$ and bases binding have rather different consequences for the structure of the bound DNA. Generally $PO_3^-$ binding stabilizes the double helix while base binding causes an unwinding of the double helix. Electrostatic forces play a major role in the stability of the double-stranded structure and the double helix can be considered as being in equilibrium under two types of opposite forces. The flexibility of a double-stranded structure will increase with the ionic strength. At a low ionic strength, the formation of an ion pair between the $PO_3^-$ and the counterion or anion and water shell play a critical role in the interaction of the metal with DNA. However, in the presence of a high ionic strength, the negative potential will vanish very close to each $PO_3^-$ group. Accordingly the anions, which were excluded at low ionic strength, can penetrate in the grooves of the double helix, together with the cations, giving the possibility for Van der Waals interaction with the bases. The changes in the melting profile of oligomers thus indicate that the binding of metal ions to DNA affects the thermal stability of the oligomers in different ways based on the metal ion's charge, ionic radii and coordination ability.

The interaction of divalent cations with DNA is extensively studied, DNA complexation with trivalent cations is less understood. It was shown that the interaction of Al with DNA produces several types of Al-DNA complexes depending on pH and metal ion concentration. However, the exact cation binding site and the effect of Al interactions on the DNA secondary structure are not determined, particularly at low cation concentrations. Martin [8] indicated that the nitrogenous bases of DNA and RNA do not bind Al(III) strongly. The weakly basic phosphate group of DNA binds Al(III) weakly, while the basic and chelating phosphate groups of nucleoside di- and triphosphates do bind Al strongly. The reasons attributed for structural changes in DNA are metal-specific binding sites on DNA (base nitrogen, phosphate oxygen). There are no definite mechanisms available in the literature to explain the role of metals in causing structural transitions.

**Acknowledgements.** We thank Department of Biotechnology, Government of India for financial support under Indo-Israel Grant on Human Genome. The authors are grateful to Dr. V. Prakash, Director, CFTRI, Mysore for his encouragement.

## 12
## References

1. Marzilli LG, Kistenmacher TJ, Eichhorn GL (1980) Nucleic acid – metal ion interactions, John Wiley and Sons, New York, USA
2. Putrament A, Baranowska H, Ejchart A, Prazmo W (1975) Mol Gen Genet 140: 339
3. Beckman RA, Mildvam AS, Loeb LA (1985) Biochemistry 24: 5810
4. Richardson CL, Verma J, Schulman GE, Shipp K, Grant AD (1981) Environ Mutagen 3: 545
5. Izatt RM, Christensen JJ, Rytting JH (1971) Chem Rev 71: 439
6. Rossetto FE, Nieboer EJ (1994) Inor Biochem 54: 167
7. Crapper DR, Krishnan SS, Quittkat S (1976) Brain 99: 67
8. Martin RB (1996) Chemistry of aluminum in the central nervous system. In: Yasui M, Strong JM, Ota K, Verity AM (eds), Mineral and Metal Neurotoxicology. CRC Press, New York
9. Jagannatha Rao KS, Rao BS, Vishnuvardhan D, Prasad KVS (1993) Biochem Biophys Acta 1172: 17
10. Rao KSJ, Rao RV, Shanmugavelu P, Menon RB (1999) Alz Rep 2: 241
11. Rogers MAM, Simon DG (1999) Age and Ageing 98: 205
12. Rao KSJ, Katsetos CD, Herman MM, Savory J (1998) Cli Lab Med 18: 687
13. Garruto RM, Yes Y (1986) Trends Neurosci 9: 386
14. Champion CSK, Rajan MT, Rao KSJ, Viswamitra MA (1998) Cell Mol Life Sci 54: 488
15. Rajan MT, Champion CSK, Vishnuvardhan D, Rao KSJ, Viswamitra MA (1996) Mol Biol Rep 22: 47
16. Karlik SJ, Eichhorn GL, Lewis PN, Crapper DR (1980) Biochemistry 19: 5991
17. Rich A (1993) Gene 135: 99
18. Anitha S, Stein R, Latha KS, Rao KSJ, Viswamitra MA (2001) Alz Rep 4: 121
19. Crapper DR, Quitkat S, de Boni U (1979) Brain 102: 483
20. Smith MA, Perry G and Richey PL (1996) Nature 382: 120
21. Lyras L, Cairns NJ, Jenner A, Jenner P, Halliwell B (1997) J Neurochem 68: 2061
22. Sugaya K, Reeves M, McKinney M (1997) Neurochem Int 31: 275
23. Yang Y, Geldmacher DS, Herrup K (2001) J Neurosci 21: 2661
24. Terry RD (1994) Prog Brain Res 101: 383
25. Lukiw WJ, Kruck TPA, McLachlan DR (1989) FEBS Lett 253: 59
26. Garner MM, Felsenfeld G (1987) J Mol Biol 196: 581
27. Herbert A, Rich A (1996) J Biol Chem 271: 11595
28. Gagna CE, Kuo Hr, Lambert WC (1999) Cell Biol Int 23: 1
29. Michak V, Maurizot MS, Charlier M (1995) J Biomol Struct Dynam 13: 565
30. Trumbore CN, Myers YN, Hyde CK (1994) Int J Radiat Biol 66: 479
31. Takeuchi H, Hanamura N, Harada I (1994) J Mol Biol 236: 610

32. Ahmad R, Naoui M, Neault JF, Diamantoglou S, Tajmir-Riahi (1996) J Biomol Struct Dynam 10: 865
33. Rao KSJ, Divakar S (1993) Bull Environ Contamn Toxicol 50: 92
34. Karlik SJ, Eichhorn GL (1989) I Inorg Biochem 37: 259
35. Walker PR, LeBlane J, Sikorska M (1989) Biochemistry 28: 3911
36. Usdin K, Grabezyk E (2000) Cell Mol Life Sci 57: 914
37. Anselmi C, Bocchinfuso G, Santis PD, Savino MAA (2000) Biophys J 79: 601
38. Negri R, Costanzo G, Buttinelli M, Venditti S, Mauro MD (1994) Biophys Chem 50: 169
39. Tassone F, Hagerman RJ, Taylor AK, Hagerman PJ (2001) J Med Genet 38: 453

# Aluminum: Interaction with Nucleotides and Nucleotidases and Analytical Aspects of Its Determination

Maria Rosa Chitolina Schetinger[1], Vera Maria Morsch[2], Denise Bohrer[3]

Departamento de Química, CCNE, Universidade Federal de Santa Maria, RS, Brazil, 97105 900
[1] *E-mail: mrosa@base.ufsm.br*
[2] *E-mail: vmorsch@base.ufsm.br*
[3] *E-mail: ndenise@quimica.ufsm.br*

Aluminum is known to affect a wide variety of biological systems. The evolution of life in a highly aluminum-rich biosphere did not result in a known physiological role of aluminum, which is considered a non-essential trace element of low toxicity in living being. In this chapter we review some toxicological and biological aspects of aluminum, as well as the interaction of this element with enzymes that use nucleotides in their reactions. It is assumed that phosphate-bearing biomolecules are elected targets for the biochemical action for aluminum. There is a large number of proteins which require nucleoside phosphates as substrates or are regulated by them. The toxicological mechanisms of aluminum in plants or animals probably involve alterations in the activity of such proteins. As examples of enzymes focus will be placed on NTPases, NTPDases, and kinases. As aluminum in biological systems occurs at trace levels, the analysis of biological matrices will be outlined with respect to aluminum determination considering contamination sources, sample collection and preparation, and methodological procedures.

**Keywords:** Aluminum, Nucleotide-binding, Nucleotidases, Contamination

## List of Abbreviations

| | |
|---|---|
| AD | Alzheimer's disease |
| ADP | adenosine 5′-diphosphate |
| Al | aluminum |
| $AlF_x$ | aluminum fluoride |
| AMP | adenosine 5′-monophosphate |
| AMS | accelerator mass spectrometry |
| ATP | adenosine 5′-triphosphate |
| ATPase | adenosine triphosphatase |
| CaM | calmodulin |
| CNS | central nervous system |
| CZE | capillary zone electrophoresis |
| DAG | diacylglycerol |
| DFO | desferrioxamine |
| DNA | deoxyribonucleic acid |
| EDTA | ethylenediaminetetraacetic acid |
| EELS | electron energy loss spectroscopy |
| ER | endoplasmic reticulum |
| ESI | electrospray ionization |
| ETAAS | electrothermal atomic absorption spectrometry |
| FDA | Food and Drug Administration |
| FPLC | fast protein liquid chromatography |
| HDPE | high density polyethylene |
| HPLC | high-performance liquid chromatography |
| IC(50) | inhibitor concentration that inhibits 50% of enzyme activity |
| ICP-MS | inductively coupled plasma mass spectrometry |
| $IP_3$ | inositol 1,4,5-triphosphate |
| *K*(i) | inhibition constant |
| LDPE | low-density polyethylene |
| NADP | nicotinamide adenine dinucleotide phosphate |
| NMR | nuclear magnetic resonance |
| NTA | nitrilotriacetic acid |

| | |
|---|---|
| PE | polyethylene |
| PI | phosphatidylinositol |
| $PIP_2$ | phosphatidylinositol 4,5-biphosphate |
| PKC | protein kinase C |
| PLC | phospholipase C |
| PP | polypropylene |
| ppb | parts per billion |
| PTFE | polytetrafluoroethylene |
| RNA | ribonucleic acid |
| TCA | trichloroacetic acid |
| UF | ultrafiltration |

## 1
## Aluminum: Toxicological and Biological Characteristics

Aluminum is the third most abundant element in the lithosphere, but its levels in natural waters, plants, and animals are relatively low [1]. It occurs in a vast variety of oxide minerals and together with silicon forms the polymorphous aluminosilicates ($Al_2O_5Si$) in rocks and soils upon which all organisms depend [2]. In soils and sediments these clays have many functions: they hold trace elements, permit growth of roots, sustain pH balance, and their cavities keep water available [3].

The evolution of life in a highly aluminum-rich biosphere did not result in a known physiological role of $Al^{3+}$, which is considered a non-essential trace element of low toxicity in healthy humans. According to various authors, aluminum levels in the human body range from 35 to 40 mg [4].

Despite its recognized toxicity, $Al^{3+}$ was considered innocuous to humans for a long time. Man extensively uses aluminum compounds, mainly as catalysts, as a reagent for water treatment, pharmaceutical products and it is often added to processed food [5]. Aluminum compounds are also used to make building materials such as bricks and cements. However, biological systems do not use this element in catalysis, do not use it to cement together the cell wall, and do not make minerals from aluminum [2, 6].

Today, aluminum is considered a toxic metal and has also been implicated in the etiology of neurological disorders such as Alzheimer's senile and presenile dementi, amyotrophic lateral sclerosis, and other diseases like aluminosis, osteodistrophy, and non-iron deficiency microcytic anemia [7–9]. $Al^{3+}$ is also considered to be the major agent causing dialysis encephalopathy in patients undergoing hemodialysis with water supplies high in aluminum [10]. In addition, aluminum is a cytotoxic metal to plants and is responsible for significant losses in world agricultural crop production and quality in acid soil, which comprises 30–40% of the world's arable lands [11]. Studies on the mechanism of $Al^{3+}$ toxicity in plants have made significant progress with several possible mechanisms having been proposed, such as cell death, inhibition of cell division, cell elongation, nutrient uptake, root elongation, ultrastructural changes in the cell wall, photosynthesis, photosynthetic electron transport, phosphorus metabolism, and ATPases activities [12–14].

Aluminum enters the human organism via the gastrointestinal (water, beverages, drugs, cosmetics, foods) or respiratory apparatus (inhaled dust), by intranasal absorption to the brain, and by transdermal absorption [15]. Most of the metal is excluded by intervention of the gastrointestinal barrier and by renal function. Also, some accumulation occurs in the entire body, including brain, heart muscle, bones, and lungs (Fig. 1).

It has been generally assumed that the amount of aluminum absorbed by healthy individuals is dependent on elimination in the urine, whereby low body levels of this element are maintained. However, it is unknown whether the normal kidney excretes all the metal absorbed when individuals are given large oral aluminum loads [16].

The absorption of aluminum depends not only on the intraluminal presence of essential metals, but also on the presence of ligands like citrate, ascorbate, gluconate, lactate, tartrate, oxalate, and malate. These dietary compounds have been shown to elevate the pH of precipitation of the aqueous $Al^{3+}$ ion from 4.5 to above 8.0 *in vitro*. These ligands probably also increase greatly aluminum absorption, both by forming low-molecular-weight luminal complexes that are better absorbed and by binding intercellular calcium, thus reducing the integrity of the tight junctions and increasing absorption through the paracellular pathway [17, 18].

To understand the role of an element in living systems we need to know not only the local accumulation and the entire amount of the element in the organism, but also the chemical forms in which it is present. Aluminum is an "A-type metal" or "hard acid" and therefore the most likely $Al^{3+}$ binding sites are oxygen donors of various biomolecules, particularly if these donors are negatively charged [19, 20]. The most important aluminum binding sites are phosphate, carboxylate, catecholate, amines, thiolates, amino acids, nucleic

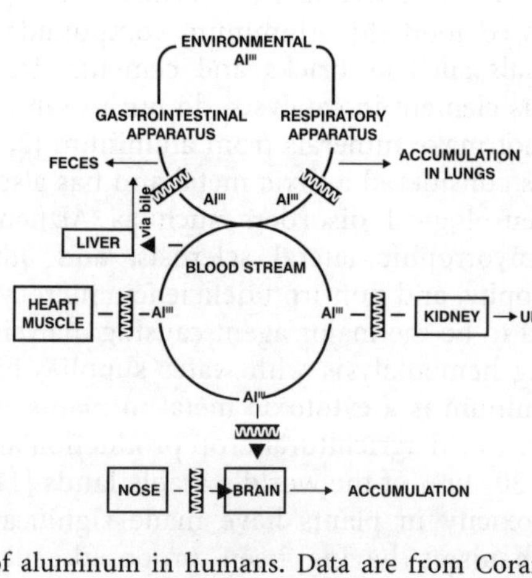


**Fig. 1.** Metabolism of aluminum in humans. Data are from Corain B, Nicolini M, Zatta P (1992) [8]

acids, and nucleotides [18]. The origin and degree of the aluminum body burden requires that metal in circulation be mobilized into excretable forms while deposited amounts are swept off by a suitable tissue/blood concentration gradient. Aluminum circulating in blood is bound to sites vacated by iron in transferrin. The aluminum transferrin complex is relatively labile, a property that makes it a possible target for sequestering ligands [7].

Williams [4] reported the formation of aluminum-containing plaques as showed below:

$$\text{(proteins)Pr} \rightarrow \text{phosphorylation}\ \ \text{PrP}_X \tag{a}$$

The higher level of phosphorylation leads to $Al^{3+}$ binding:

$$\text{PrP}_X + Al^{3+} \rightarrow \text{PrP}_X \cdot Al^{3+} \tag{b}$$

The $PrP_x \cdot Al^{3+}$ complex can cause cell death and also promotes the formation of some plaques. There is a possibility of further reaction with some cations, normally present in high concentration inside and outside the cell:

$$\text{PrP}_X Al^{3+} + M \rightarrow \text{PrP}_X AlM \downarrow \text{(plaque)} \tag{c}$$

## 1.1 Aluminum Chemistry

Aluminum bioavailability depends on the chemistry of the $Al^{3+}$ ion *in vivo*. In particular, the high charge to radius ratio of this element, which determines its specific attraction for oxygens donors, is at the origin of its deleterious effects on macromolecular biological structures. In this context a thorough understanding of aluminum hydrolysis is essential before we can go on to characterize accurately other ligand systems [7, 9, 21]. Biological systems constantly contain many potential ligands, and the formations of ternary complexes may be very important. $Al^{3+}$ has a strong hydrolytic tendency, so ternary Al-ligand-hydroxide complexes represent most of the ligands at physiological pH.

**Table 1.** Ligands and constants $K_D$** for aluminum complexes

| Ligand | Complex | $-\text{Log}\ K_D$ |
|---|---|---|
| Citrate (Cit) | $Al(Cit)^0$ | 8.1 |
| Desferroxiamine | Al(Df) | 24.1 |
| EDTA | $Al(EDTA)^-$ | 16.2 |
| Fluoride | $AlF^{2+}$ | 6.2 |
| Hydroxide | $Al(OH)^{2+}$ | 8.5 |
| Phosphate | $Al(HPO_4)^+$ | 9.7 |
| Transferrin | AlT | 13.0 |

* Adapted from Williams (1999) [2].

** $K_D$ is the true dissociation constant (pH = 7.0).

A large variety of anions has high affinity for aluminum in aqueous solution [2]. Table 1 lists some of the most important ligands for this element.

The $Al^{3+}$ ion shows a propensity to suffer hydrolysis which can produce the formation of various mononuclear hydroxo complexes. At pH around 7.0 it often precipitates as $Al(OH)_3$, which redissolves leading to the formation of aluminate, $Al(OH)_4^-$. After this, polynuclear complexes such as $Al_2(OH)_2$, $Al_3(OH)_4$, and $Al_{13}(OH)_{32}$ can also form. These species can produce, in a very slow process, oligomeric complexes, which are necessary precursors of the macromolecular polymer $[Al(OH)_3]_n$ [16, 17]. In Fig. 2 we can observe the distribution of the $Al^{3+}$ aquo ion and the species for 100 mM and 10 μM total aluminum. The maximum concentration of $Al(OH)_3$ is reached at about neutral pH. At pH ≥ 7.0 the predominant species is aluminate ion $Al(OH)_4^-$.

Phosphate is a fundamental nutrient to all living systems and also a ligand of considerable physiological importance that forms complexes with aluminum. Inorganic and organic phosphates are important constituents of both extracellular and intracellular fluids in the human body (around 1–10 mM total phosphate) [22]. Öhman and Martin [23] observed that, although frequently denominated as $AlPO_4$, this species appears to contain variable proportions of phosphate and hydroxide in a mixed phosphate-hydroxo complex. The anion $PO_4^{3-}$ does not exist in aqueous solution at pH values under 12, and at physiological pH values, $HPO_4^{2-}$ is the dominant species. $Al^{3+}$ can form the following salts with inorganic phosphate: monoaluminum phosphate – $Al(H_2PO_4)_3$, aluminum metaphosphate – $[Al(PO_3)_3]_n$ and sodium aluminum phosphate – $NaAl_3H_{14}(PO_4)_8 \cdot 4\,H_2O$, $NaAl_3H_{14}(PO_4)_8$ and $Na_8Al_2(OH)_2(PO_4)_4$. The orthophosphate anion is the basic anionic unit of all phosphates. The anion possess three valences that can be fulfilled by hydrogen, metal ions or a combination of both. The tetrahedron structure allows the formation of polymers with one, two or

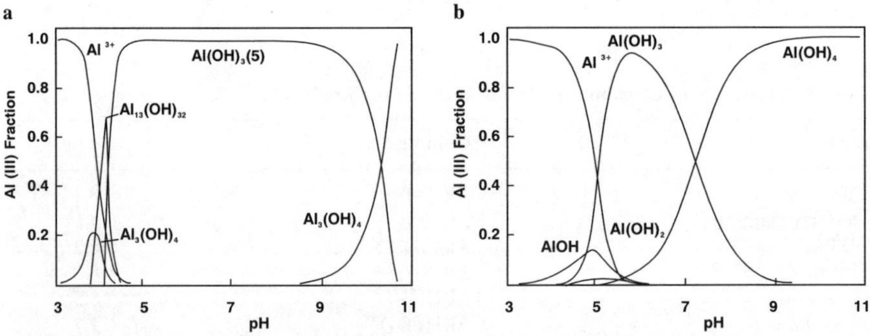


**Fig. 2.** Distribution of hydrolytic species formed in aqueous solution of Al(III) ion as a function of pH. Total Al(III) concentrations: (**a**) 100 mM; (**b**) 10 μM. Data are from Kiss T (1995) [9]

three units of the phosphate anion where each unit is linked by shared oxygen atoms [22].

Organic phosphates may be divided into 2 classes: basic and weakly basic phosphates. The first class is monosubstituted with p*K* between 6–7. It possesses the formula R-$OPO_3^{2-}$ and is present as the terminus phosphate in nucleoside mono-, di-, and triphosphates. The weak phosphates are disubstituted and show the formula R-O(R′-O)$PO_2^-$, p*K*$<$2 and occur as the inside phosphates in nucleoside di- and triphosphates and in nucleic acids DNA and RNA [16–18, 24, 25].

The larger biomolecules that have potential $Al^{3+}$ binding sites are phosphate-bearing biomolecules such as ATP, membrane phospholipids and nucleic acids. It is important to recognize that the metal coordination of these biomolecules might lead to serious disgorges in central biological processes necessary to cell homeostasis and consequently for its overall healthy condition [21].

Adenosine 5′-triphosphate (ATP) is a well recognized molecule of importance as the major carrier of chemical energy in the cells. It is a major extracellular signaling molecule all over the body. ATP is a nucleotide formed by an adenine, a ribose, and a triphosphate unit and exists in the cell as a complex with $Mg^{2+}$ [26]. Aluminum can replace magnesium in many biological processes by forming a more stable complex with ATP than $Mg^{2+}$ can form. At a cellular level, high aluminum concentrations suppress the phosphorylation, resulting in a reduction of blood adenosine triphosphate/adenosine diphosphate (ATP/ADP) ratio and the level of ATP. A large number of proteins requires ATP as substrates or is regulated by it. So, it is important to study the interaction among the elements that can interfere in the equilibrium with these nucleotide [27].

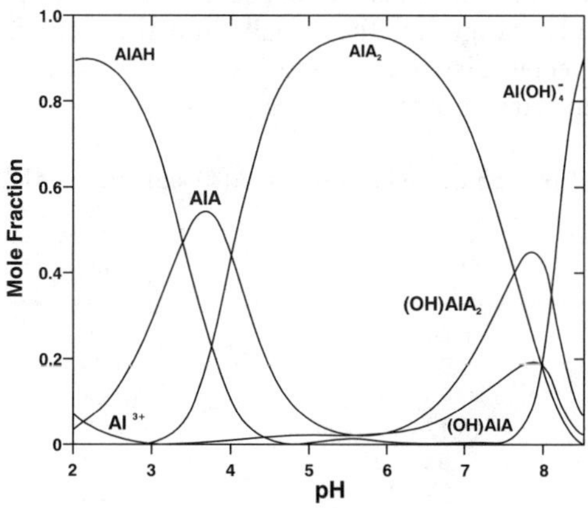


**Fig. 3.** Species distribution curves for 0.1 mM total aluminum and 10 mM total ATP. The constants are referred in Table 2. A refers to tetraanionic $ATP^{4-}$. Data are from Nelson DJ (1996) [18]

Aluminum binds to nucleoside phosphates mainly through the basic terminal phosphate groups. Nucleosides mono-, di-, and triphosphates demonstrate similar phosphate basicity. Aqueous solutions of $Al^{3+}$ and nucleoside phosphates have a tendency to form ternary complexes with hydroxide in a pH-dependent manner. In addition, there is a possibility of $Al^{3+}$-bridged complexes being formed. Fig. 3 shows the species distribution for the $Al^{3+}$-ATP system. At physiological pH the merged hydroxo mono complexes predominate [9, 18].

In Table 2 we can observe that the first stability constant (log $K_1$) for $Al^{3+}$ binding to all deprotonated phosphate groups decreases through the series ATP $\rightarrow$ ADP $\rightarrow$ AMP (7.92, 7.82, and 6.17) [9]. Jackson and Voyi [28] made a study on the chelation of aluminum with AMP, ADP, and ATP in biological systems and by using potentiometry they detected that the complexes formed with this element and nucleotides are fairly stable. They attributed the toxicological effect of aluminum to coordination to one of these high-energy phosphate compounds.

In addition to ATP, there are other phosphate-bearing biomolecules that play roles in human physiology: ADP, AMP, 2,3-diphosphoglycerate, inositol phosphate, glucose 6-phosphate, etc. [7, 9, 21].

Champmartin et al. [29] studied the interaction of $Al^{3+}$ with glucose 6-phosphate in aqueous solution using pH-potentiometry and multinuclear ($^{31}P$-, $^{27}Al$-, $^{13}C$-) NMR spectroscopy. The pH range tested was 1 to 8 because above this value the formation of precipitates occurs. They verified the formation of mononuclear species such as $MLH_2$, MLH, ML, $ML_2H$, $ML_2$ and $MLH_{-3}$, and also dinuclear complexes like $M_2L_2H_{-n}$ (n = 1–4) (Fig. 4).

Venturini-Soriano and Berthon [30] carried out a quantitative investigation of $Al^{3+}$-malate complex in biological fluids. Using potentiometric titrations they established a correlation between the character of malate as a ligand and the distribution of aluminum in the gastrointestinal fluid. According to the authors, malate shows to be extremely capable of maintaining $Al(OH)_3$ soluble over the entire pH range of the small intestine under normal dietary conditions. The species distribution curves of $Al^{3+}$-malate complexes in the gastrointestinal fluid are shown in Fig. 5.

**Table 2.** Adenosine 5′-phosphate acidity constants and their $Al^{3+}$ complex stability constants*

| Parameter | AMP | ADP | ATP |
|---|---|---|---|
| HA | 6.04 | 6.19 | 6.31 |
| $pK_a$ | 3.74 | 3.79 | 3.89 |
| AlAH | – | 10.98 | 11.30 |
| AlH (log $K_1$) | 6.17 | 7.82 | 7.92 |
| (OH)AlA | 2.02 | 2.94 | 2.46 |
| $AlA_2$ | 10.35 | 12.16 | 12.47 |
| $(OH)AlA_2$ | Precipitation | 5.01 | 4.84 |

Table taken from Kiss et al. (1996) [17]. Constants are given as log at 25 °C and 0.2 M ionic strength controlled with KCl. Acidity constants are concentration constants.

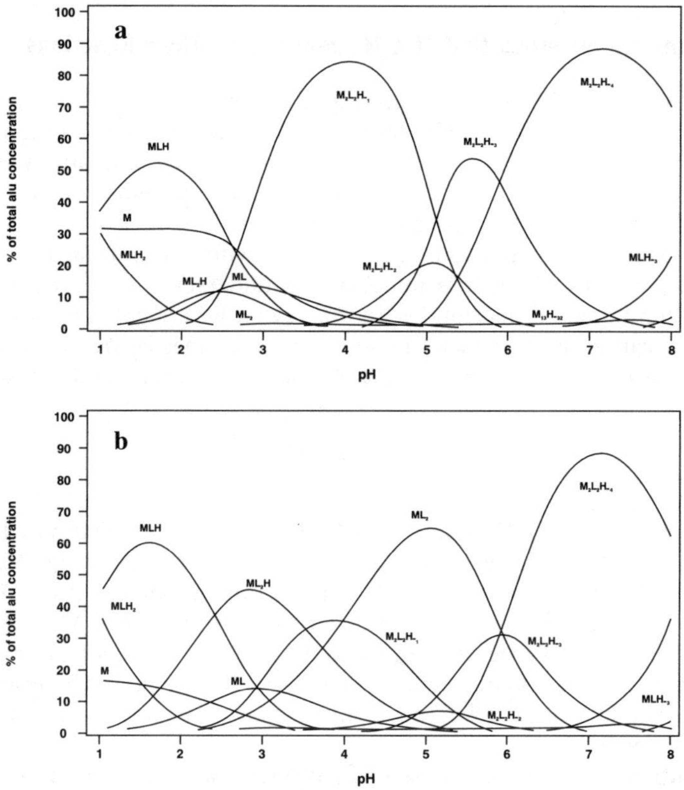


**Fig. 4.** Species distribution in the $Al^{3+}$-Glucose 6-phosphate system for a ligand to metal (L/M) ratio of 1:1 (**panel A**) and 3:1 (**panel B**); T = 298 K; $Al^{3+}$ = $2 \times 10^{-2}$ (**A**) and $6.7 \times 10^{3}$ (**B**) M. Data are from Champmartin D, Rubini P, Lakatos T, Kiss T (2001) [29]

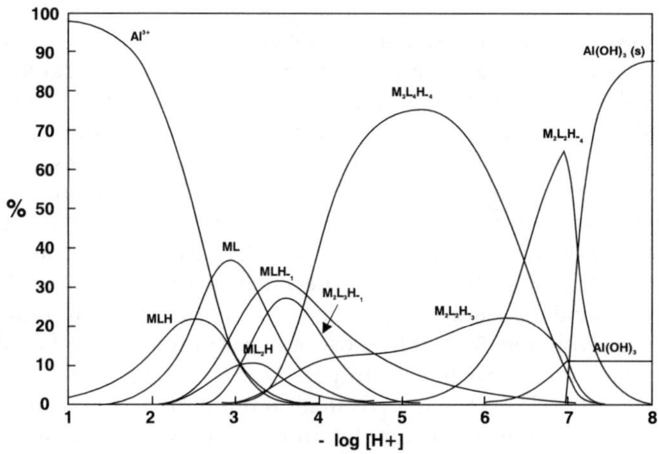


**Fig. 5.** Distribution profiles for 0.0005 M aluminum and 0.01 M malate in the gastrointestinal fluids. Data are from Venturini-Soriano M, Berthon G. (2001) [30]

## 2 Aluminum and Enzymes that Use Nucleotides in Their Reactions

The molecular mechanisms underlying the detrimental effects caused by aluminum have yet to be fully understood [31]. The toxicological mechanisms of aluminum in plants or in animals probably involve alterations in many enzyme activities. There is a large number of proteins which requires nucleoside phosphates as substrates or is regulated by nucleoside phosphates, therefore aluminum can potentially interfere with the normal functions of these proteins [18]. Thus, it is important to understand some reactions that involve aluminum and nucleotides. In this review we are going to focus on the alterations caused by this metal in some enzymes that hydrolyze or transfer phosphate groups of nucleotides, like NTPases, NTPDases, and kinases.

### 2.1 NTPases (ATPases, GTPases)

The enzymes called ATP phosphohydrolase are widely distributed in the evolutionary chain and in biological systems. In some cases the ATPase is activated either by magnesium ($Mg^{2+}$ ATPase) or by calcium ($Ca^{2+}$ ATPase), and in other cases by both calcium and magnesium ($Ca^{2+}$ $Mg^{2+}$ ATPase). Another class of ATPase is stimulated by sodium and potassium and is inhibited by ouabain being denominated $Na^{+}$ $K^{+}$ ATPase. There are some ATPases that hydrolyze other nucleotides than ATP, however, with a high preference for ATP.

In addition, there are also a class of enzymes that hydrolyses preferentially the nucleotide GTP being denominated GTP phosphohydrolase or GTPase.

#### 2.1.1 *$Ca^{2+}$ ATPases*

The first enzyme to be discussed in this review is $Ca^{2+}$ ATPase. $Ca^{2+}$ ATPase, a calmodulin (CaM)-regulated membrane-bound enzyme is responsible for expelling calcium from the cells and thus maintaining low intracellular calcium levels. Some data reinforce the hypothesis that aluminum is either directly or indirectly able to disrupt calcium homeostasis, with many pathological consequences [31, 32]. The effect of aluminum (10 mg $kg^{-1}$ body weight $day^{-1}$ i.p.) for a period of 4 weeks was demonstrated to be a detrimental action on the activity of $Ca^{2+}$ ATPase in the central nervous system (CNS) (Fig. 6). In the *in vitro* experiments (using 0–120 μM $Al^{3+}$) Julka and Gill [32] found a dose-dependent decrease in the activity of synaptosomal $Ca^{2+}$ ATPase, with the 50% inhibitory concentration ($IC_{50}$) for aluminum being calculated as approximately 10 μM. Corroborating with these findings, Gandolfi et al. [31] demonstrated that aluminum acts as an important disrupter of intracellular calcium by interfering with different regulatory systems such as those of the plasma and endoplasmatic reticulum (ER) $Ca^{2+}$ ATPase (Fig. 7) and mitochondrial ATPase. The authors observed that

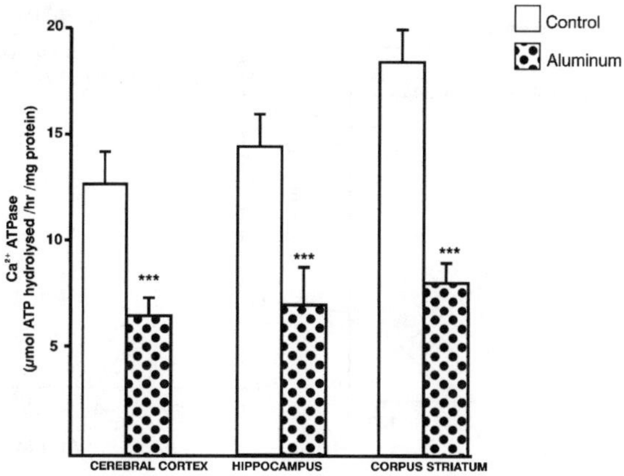


**Fig. 6.** Effect of *in vivo* aluminum exposure on the $Ca^{2+}$ ATPase activity in synaptic plasma membranes. Values are mean ± SD of 8 animals/group. *** $P < 0.001$, statistically significant from control group. Data are from Julka and Gill (1996) [32]

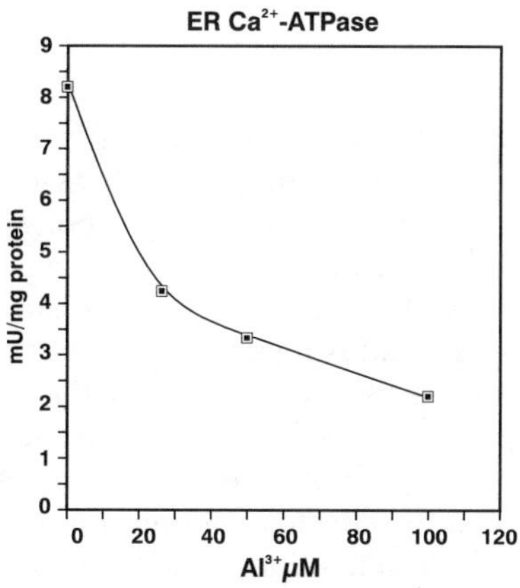


**Fig. 7.** Effect of $Al^{3+}$ on ER calcium-ATPase, reported as enzyme activity (mU/mg protein) vs. metal ion concentration (µM). The curve exhibits an inhibitory saturation type phenomenon dependent on Al concentration. Data represent the average of three independent experiments carried out in triplicate. Data are from Gandolfi et al. (1998) [31]

the inhibition of rat liver ER $Ca^{2+}$ ATPase (ER $Ca^{2+}$ATPase catalyses an active transport of calcium across the ER membrane) represents a saturation-type phenomenon dependent on aluminum concentration (Fig. 7). On the other hand, the effects of aluminum on $Ca^{2+}$ ATPase from rat brain or rat cerebellum homogenates were dose-dependent (Fig. 8). One of the major consequences in the increase of intracellular calcium levels is the enhanced production of free radicals, which can have damaging effects on the integrity of cellular membranes [32].

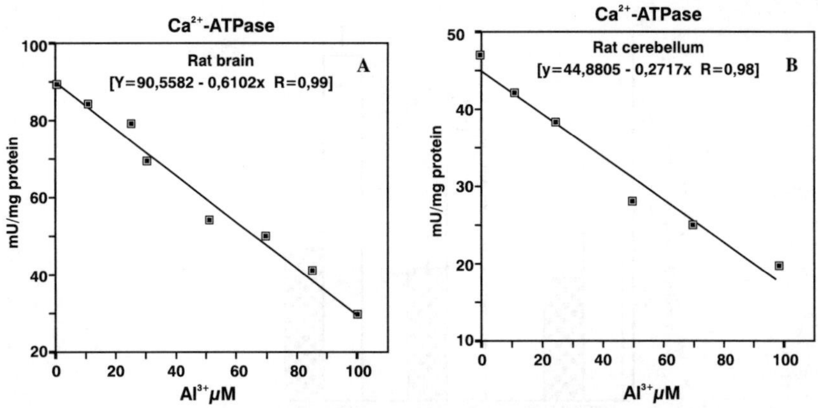


**Fig. 8.** Effect of $Al^{3+}$ on calcium ATPase activity (mU/mg protein) from homogenate of rat brain (**panel A**) and cerebellum (**panel B**). The inhibition by $Al^{3+}$ is dose-dependent; data represent the average of three independent experiments carried out in triplicate. Data are from Gandolfi et al. (1998) [31]

### 2.1.2
### *$Na^+$ $K^+$ ATPase*

The second enzyme to be discussed is $Na^+$ $K^+$ ATPase. $Na^+$ $K^+$ ATPase or the sodium-potassium-activated adenosine triphosphatase has been considered to be an integral part of the sodium pump and is responsible for the transport of potassium into and sodium out of a variety of cell types. It is present at high concentrations in brain cellular membrane, comprising about 40–50% of the ATP generated in this tissue, and is highly responsive to changes in membrane fluidity [33]. This activity is very important in the mechanism involved in conduction and synaptic transmission in the nervous system and helps to maintain the osmotic regulation besides other important functions in the cells. Some authors have studied the effects of aluminum on $Na^+$ $K^+$ ATPase. In the work of Zatta et al. [34] an activation on $Na^+$ $K^+$ ATPase of the rat brain homogenate as a dose/effect-dependent phenomenon was observed (Fig. 9, *panel A*). They observed an activation of rat liver mitochondrial $Na^+$ $K^+$ ATPase (Fig. 9, *panel B*) and suggested that the opposite results, obtained by other authors and describing an inhibitory effect of aluminum, are probably due to the utilization of ill-defined toxicological protocols. In fact, in a previous work of 1980 [35], the inhibitory effect of $AlCl_3$ on $Na^+$ $K^+$ ATPase was observed with a very high IC(50) of 8.3 mM, with aluminum being considered a very poor inhibitor of such enzymes [35].

### 2.1.3
### *$Mg^{2+}$ ATPase*

In this sequence we will now discuss $Mg^{2+}$ ATPase. As previously cited, aluminum replaces magnesium in some reactions. Therefore aluminum is

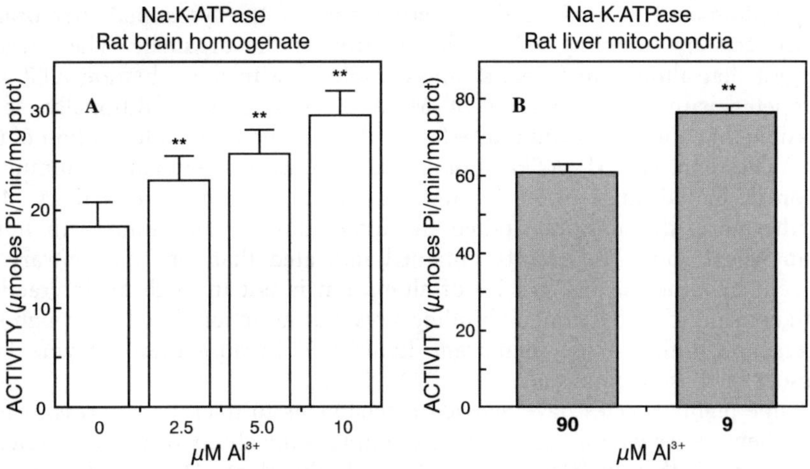


**Fig. 9.** (**Panel A**) ATPase activity assay in the rat homogenate expressed as μMol $min^{-1}$ $mg^{-1}$ protein, either in the controls or in the presence of different concentrations of Al(III). Bars ± SD $^{**}P < 0.01$. (**Panel B**) ATPase activity assay in rat liver mitochondria expressed as μMol $min^{-1}$ $mg^{-1}$ protein, in the controls or in presence of 9 μM Al(III). Bars ± SD. ** $P < 0.01$. Data are from Zatta et al. (1995) [34]

likely to interfere with normal functions of enzymes that require magnesium (usually with inhibition) [18]. Martin [36] and Macdonald et al. [37] pointed out that the inhibition of these enzymes by aluminum could have both kinetic and thermodynamic bases.

The mitochondrial $Mg^{2+}$ ATPase activity from liver and brain was inhibited by about 50–60% after an administration of 100 mg Al $kg^{-1}$ body weight in the diet for a period of 90–120 days. By contrast, in the heart mitochondria, the ATPase activity increased from 73 to 212% after this treatment [38]. In this work, it was also found that ADP phosphorylation rates were decreased by 46% and that the changes in the ATPase activity, in general, were paralleled to those of the respiratory rates. The author suggested that these results imply that the effects of $Al^{3+}$ treatment on respiratory activity and the ATPase activity go hand in hand. Curiously, liver and brain mitochondria presented doubled aluminum concentration and impaired respiration rate, whereas the heart mitochondria, that accumulated 11 times higher amount of aluminum, presented stimulation of respiration. Thus, an indirect action of aluminum in this tissue could be suggested.

In certain human pathologies, as well as in experimental animals, aluminum has been observed to accumulate preferentially inside lysosomes. The experimental evidence raised the possibility that lysosomes serve as an elective site of metal accumulation. Lysosomes might thus protect the rest of the cell from toxic effects produced by free metals or metal compounds. As reported by Zatta et al. [39] aluminum is able to inhibit the lysosomal $H^+$ ATPase activity from rat liver. Intralysosomal pH is acidic, and it is accepted that it is maintained by the proton pump which is an enzyme activated by magnesium and located in the external part of the lysosomal membrane, which acts by pumping $H^+$ to the interior of the organelle. The authors suggest that aluminum forms strong complexes with the substrate ATP and competes with the formation of Mg-ATP complexes. Additionally, it is possible that aluminum could interfere with the structural conformation of the $H^+$ ATPase. In plant $H^+$ ATPase, obtained from wheat microsomes, aluminum chloride, in the range of 50–200 μM, was unable to inhibit its activity [40]. Furthermore, the authors studied the interaction of aluminum with lipids from wheat root. The results obtained indicated that, on the contrary to current hypotheses, the toxicity of aluminum is not through an interaction with enzymatic catalytic metal binding sites, but its action may be through the interaction with specific membrane lipids [40] because these enzymes are present in these membranes.

Some plants species have evolved mechanisms to tolerate Al stress, which help them to grow on acid soils. One example is the ability of plants to release organic acids that chelate aluminum outside the plant [41]. A wide range of proteins is induced by aluminum stress in wheat and other plant species, including membrane-bound proteins [42, 43]. The RMP51 (root membrane protein with 51 kD) is expressed in an Al-resistant cultivar (cv) of wheat, cv PT741 and it was demonstrated that the RMP51 band consists of the $\beta$-subunit of the V-ATPase and the $\alpha$-and $\beta$-subunits of the $F_1F_0$-ATPase. V-ATPase activity increased by 1.6- to 2.1-times in the presence of 0 to 100 μM $AlCl_3$, while the levels of the transcript encoding the $\beta$-subunit increased 2.2-times. Mitochondrial $F_1F_0$-ATPase activity increased by 2.1- to 7.3-times with a constant level of the transcript encoding the $\alpha$-subunit over the same range of Al concentrations [42]. It was suggested by the authors that the induction of the V-ATPase and the increase in the activity of $F_1F_0$-ATPase play a role in Al tolerance in the cultivar PT741.

An inhibition of an $Mg^{2+}$ ATPase on corn root plasma membrane was also observed. Kinetic data on aluminum inhibition present a competitive pattern, as demonstrated by the Lineweaver-Burk plot with an apparent inhibition constant ($K_i$) of 40 μM [44]. These results were obtained at pH 6.6. The authors suggested that the inhibition may be a result of either the formation of an inefficient substrate (Al-ATP) or an interaction directly with the enzyme structure.

In the alga *Chlorella vulgaris*, a pronounced inhibition of $Mg^{2+}$ ATPase and $Ca^{2+}$ ATPase occurred by the interaction of $AlCl_3$ + NaF. These effects were additive at pH 6.8 and synergistic at pH 6.0 and 4.5 [14]. Thus, the authors suggest that the inhibition could be due to interference with the phosphate binding site of ATPases, thereby arresting the release of energy.

### 2.1.4
### *$Ca^{2+}$ $Mg^{2+}$ ATPase*

The ecto-nucleotidase responsible for the hydrolysis of ATP to ADP in the coronary endothelium is activated by both $Ca^{2+}$ and $Mg^{2+}$ and is described as a $Ca^{2+}$ $Mg^{2+}$ ATPase. In the work of Korchazhkina et al. [45] a high performance

**Table 3.** The influence of Al in the hydrolysis of two different concentrations of ATP[a]

| [ATP] nominal (μM) | ATP hydrolyzed (nMol $min^{-1}$ $g^{-1}$) | | |
|---|---|---|---|
| | −Al | +Al | |
| 1.0 | 29.5 ± 3.14 | 31.7 ± 2.80 | $P = 0.622$ |
| 5.0 | 135.1 ± 8.84 | 129.6 ± 10.0 | $P = 0.543$ |

[a] The concentration of metabolites measured are presented as nMol $min^{-1}$ $g^{-1}$ to account for differences in coronary flow and weight of individual hearts. Mean and SEM. are given, n = 5. Data are from Korchazhkina et al. (1999) [45].

liquid chromatography (HPLC) method was used for testing the hydrolysis of ATP in the coronary circulation of the isolated working rat. The method allows the simultaneous determination of ATP and the products of its hydrolysis in the coronary effluent of an isolated working rat heart. The results obtained demonstrated that $Ca^{2+}$ $Mg^{2+}$ ATPase did not alter ATP hydrolysis in the presence of an excess of aluminum (Table 3). Furthermore, the further breakdown of ADP to AMP and AMP to adenosine was efficient and similarly unaffected by the presence of aluminum.

### 2.1.5
### *GTPase*

Guanine nucleotides like GTP are common regulators of signal transduction process and can exert their effects through direct interaction with an effector system or via intermediary phosphorylation events [46]. G-proteins (guanosine nucleotide-binding proteins) are activated by Mg-dependent GDP/GTP exchange. An intrinsic GTPase activity restores the resting GDP state. Aluminum displacement of endogenous $Mg^{2+}$ has been demonstrated on the trimeric G-protein, a transducin with GTPase activity. As a result of aluminum binding to the metal · GTP · transducin complex, the rate of the $Mg^{2+}$-dependent GDP/GTP exchange reaction was inhibited by 25–60% [47]. Inhibition of G-protein activation by aluminum is attributed to the enhanced thermodynamic stability of the Al · GDP · protein ternary complex relative to the Mg · GDP · protein complex [48]. Macdonald et al. [37] demonstrated an inhibition by aluminum on the GTPase activity of the microtubules. More recently, Landino and Macdonald [49] demonstrated that picomolar aluminum concentrations inhibited GTPase activity of ras p21 in an Mg-dependent manner (Table 4), consistent with an Al/Mg competition mechanism. The affinity constant for aluminum on the metal · GDP/GTP · p21 ternary complex is five to six orders of magnitude greater than for the corresponding $Mg^{2+}$ complex. These findings corroborate the idea of Al/Mg substitution as a potential biochemical mechanism for aluminum toxicity [48].

Another approach focuses on the effects of aluminum fluoride anions on protein G. The effect of aluminum on signal transduction in the absence and presence of fluoride ion is entirely distinct. In the absence of fluoride ion,

**Table 4.** Reversion by $Mg^{2+}$ of the inhibition of *ras* p21 GTPase activity caused by $Al^{3+}$

| $[Mg]^{2+}$. (μM) | % GTPase activity | $[Mg^{2+}]/[Al^{3+}]$ |
|---|---|---|
| 10 | 13 ± 1 | $3.5 \times 10^4$ |
| 100 | 42 ± 2 | $3.5 \times 10^5$ |
| 1000 | 88 ± 4 | $3.5 \times 10^6$ |

Reactions contained 10–30 nM $Mg^{2+}$ · GDP · p21, the designated $Mg^{2+}$concentrations, $2.9 \times 10^{-10}$ M $Al^{3+}$, and μM [$\gamma$-$^{32}$P]-GTP in 20 mM Tris-acetate pH 6.9, 100 mM NaCl, 1 mM DTT. GTPase activity was measured by monitoring the release of $^{32}$Pi from [$\gamma$-$^{32}$P]-GTP for 5 min. % GTPase activity is (+$Al^{3+}$/NTA)/(NTA control) at each $Mg^{2+}$ concentration. Data are from Landino and Macdonald (1997) [49]. NTA: nitriloacetic acid.

aluminum causes overall signal transduction to decrease (because less GTP binds) while, in the presence of fluoride ion, $Al^{3+}$ markedly enhances signal transduction [18]. Aluminum fluoride has been suggested to stimulate G-protein, at the nucleotide binding site in the $\alpha$-subunit, by forming a GDP-$Al^{3+}$-$F^{3-}$ complex that imitates GTP [50], which can potentially lead to a dysfunction in biological systems [18]. In this effect, $AlF_x^-$ (where x is probably 3) mimics the $\gamma$-phosphate group of GTP to stabilize the G$\alpha$ subunit of the regulatory protein in the transitional state [51]. It was suggested that the $AlF_x^-$ mechanism is linked to the phosphatidylinositol (PI) system [52]. The receptor-stimulated hydrolysis of phosphatidylinositol 4,5-biphosphate ($PIP_2$) to inositol 1,4,5-triphosphate ($IP_3$) and diacylglycerol (DAG) by phospholipase C is an important transmembrane signaling systems in neurons and in other cells. The generation of $IP_3$ and DAG causes the release of calcium from intracellular stores and stimulates kinases [53]. Confocal imaging of intracellular $Ca^{2+}$ clearly demonstrates that $AlF_3^-$ could increase the internal $Ca^{2+}$ concentration in oocytes in the absence of external $Ca^{2+}$ [54] by activating G-proteins. In another work, Matzel et al. [46] suggested that $AlF_3^-$ may directly supplement the calcium load produced by depolarization by enhancing the current flux through voltage-dependent calcium channels or by releasing calcium from intracellular stores through stimulation of the phospholipase C pathway. In addition, in another receptor-mediated signal transduction pathway, $AlF_4^-$ acts as a competitive inhibitor. In cabbage, aluminum inhibits the enzyme phospholipase D (that hydrolyses the terminal phosphodiester bond of membrane phospholipids, mainly phosphatidylcholine), by occupying a phosphate-binding locus on the active site to prevent the formation of a phosphatidyl-enzyme intermediate [55].

Besides these effects of $AlF_3^-$ on G protein systems, aluminum has its own actions on the phosphoinositide signaling pathway. Aluminum specifically inhibits the $Ca^{2+}$-dependent enzyme phospholipase C which acts on $PIP_2$. It was found that aluminum chloride inhibited the hydrolysis of $PIP_2$ in a concentration-dependent manner with an IC(50) slightly above 100 μM [56]. The inhibition observed is competitive in nature with the substrate $PIP_2$ [57].

## 2.2 NTPDase

NTPDases (apyrase, ATP diphosphohydrolase, ecto-ATPDase) are ubiquitous enzymes with a low substrate specificity. These enzymes had previously been identified in biochemical terms and were shown to hydrolyze a variety of purine and pyrimidines nucleoside di-and triphosphates [58]. NTPDases are normally activated by divalent cations, preferentially calcium and magnesium, having an alkaline pH optimum and are not inhibited by typical inhibitors of ATPases such as vanadate, ouabain, and low azide concentrations. The synaptosomal NTPDase from rat cerebral cortex was inhibited by aluminum chloride, used in the range of 0.01 to 10 mM, in the presence of calcium as demonstrated by Schetinger et al. [59]. It is important to pinpoint that the aluminum inhibition was competitive with respect to calcium for NTPDase with both ATP and ADP as substrates (Fig. 10). The results obtained suggested that aluminum forms complexes with ATP and ADP and can be recognized by the active site, however they are not such good substrates for the enzyme as the calcium-nucleotide complexes. In the central nervous system, ATP acts as a neurotransmitter in the purinergic system, by acting on purinoreceptors [60]. These results suggested that aluminum impairs the degradation of the neurotransmitter ATP in the cerebral cortex. Hence, a possible over-stimulation of the $P_2$ purinoreceptors can be predicted and the actions are activated via, or perhaps by a desensitization of, these receptors in the presence of aluminum.

## 2.3 Kinases

Previously, the enzymes that hydrolyze nucleotides were discussed, now we will discuss the enzymes that transfer phosphate groups. Kinases are enzymes that catalyze the phosphorylation by transferring the $\gamma$-phosphate of ATP to specific substrates.

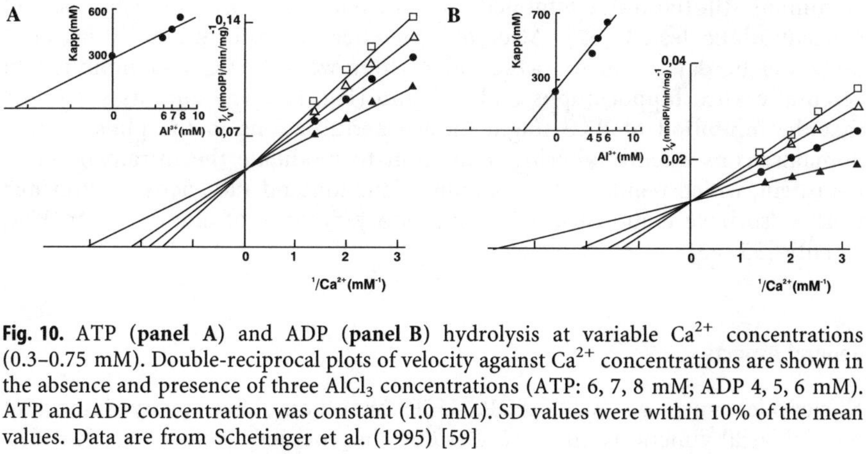


**Fig. 10.** ATP (**panel A**) and ADP (**panel B**) hydrolysis at variable $Ca^{2+}$ concentrations (0.3–0.75 mM). Double-reciprocal plots of velocity against $Ca^{2+}$ concentrations are shown in the absence and presence of three $AlCl_3$ concentrations (ATP: 6, 7, 8 mM; ADP 4, 5, 6 mM). ATP and ADP concentration was constant (1.0 mM). SD values were within 10% of the mean values. Data are from Schetinger et al. (1995) [59]

### 2.3.1
### *Hexokinase*

The inhibition by aluminum on the activity of the glycolytic enzyme hexokinase, that catalyzes the phosphorylation of glucose or fructose in the presence of $Mg^{2+}$ ATP, was demonstrated by Exley et al. [61]. In addition, in this work it was also reported that the biological availability of aluminum was dependent upon the equilibrium state of the aluminum stock solution. The aluminum stock solutions that, prior to their addition to the assay, had been preequilibrated for 24 h as opposed to only 1 h, showed an enhanced inhibition of the assay activity rate. The authors suggested that the inhibition of hexokinase be probably due to the fact that ATP binds aluminum to form the complex ATP-Al, in preference to its usual cofactor magnesium to form the complex ATP-Mg. Thus, there is a competition between these two complexes formed in the active site and probably aluminum does not bind solely to hexokinase. In plants, a substantial portion of the hexose phosphorylation activity is associated with mitochondria [62]. Mitochondria isolated with a high degree of intactness from avocado mesocarp, presented an associated hexokinase with a high preference for glucose as the hexose, and Mg-ATP as the most effective phosphoryl donor. In this preparation, when the pH was 6.8, aluminum showed to be a potent inhibitor. However, when the pH was 8.2, no inhibition was found and at pH 6.8, 50% inhibition of hexokinase was estimated to occur at 12 μM of aluminum.

### 2.3.2
### *Protein Kinase C*

Protein Kinase C (PKC) is a $Ca^{2+}$ phospholipid-dependent kinase. It is highly concentrated in the brain, localized in the presynaptic nerve terminal and regulates neurotransmitter release in the central nervous system via a G protein-coupled mechanism. Incubation with varying concentrations of aluminum (0–100 μM) inhibited the activity of PKC of CNS at all concentrations tested [32]. Also, *in vivo* after an exposure of 10 mg $kg^{-1}$ body weight $day^{-1}$ i.p. for a period of four weeks, PKC was inhibited in cerebral cortex, hippocampus and striatum (Fig. 11). The authors suggested that the inhibition of PKC might induce perturbations in the phospholipid domains, consequently altering aluminum to modulate the intramembranal G-protein, which regulates $Ca^{2+}$ channel function and PKC activity. Previous works also have demonstrated the inhibitory potency of aluminum on PKC activity [55, 63].

## 2.4
## General Aspects

Nucleotides such as ATP, ADP, UTP or others have many important physiological functions in cells. Besides the energetic role of nucleotides,

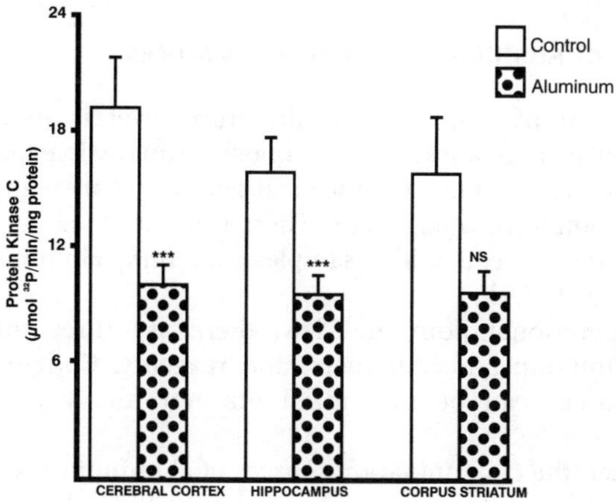


**Fig. 11.** Effect of *in vivo* aluminum exposure on calcium regulatory protein, PKC. Values are mean ± S.D. of 8 animals/group. $*P < 0.05$, $***P < 0.001$, statistically significant from control group

these molecules are signaling substances in many tissues. It is important to pinpoint that aluminum, as a toxicant, affects a wide range of enzymes activities either with inhibition or activation. In whole-cell systems, the intracellular effects of an extracellular $Al^{3+}$ challenge are often biphasic, being stimulatory at low concentrations and inhibitory at higher concentrations [64]. It could be explained as a U-shaped dose-response continuum being stimulatory at low doses and inhibitory at high doses, leading to the biphasic, hormetic dose-response curve that occurs as a response to a disruption in homeostasis [65]. The explanation is that, at low doses of disruption or toxicity, many biological systems display an overcompensation response, which results in the apparent low-dose stimulation component of the response curve. At higher doses with greater initial toxicity, the system often displays a more limited capacity for a compensatory response, usually insufficient to return to control levels. These opposite effects can disrupt the normal biochemical and physiological pathway where these enzymes are located. Aluminum may inhibit or activate by binding at a site other than a metal ion binding site, such as one or more non-phosphorylated or phosphorylated amino acid side chains, or to a substrate-cofactor. Also, in some instances the aluminum response is linked to the presence of an additional non-protein ligand, such as fluoride [18].

At this point, it is clear that aluminum can interact with phosphate groups of nucleotides or in enzyme catalytic structures, thus it is probable that many other enzymatic reactions not discussed here could be damaged by aluminum.

# 3
# Determination of Aluminum in Biological Matrices

The determination of aluminum at the trace level is more than simple laboratory experimental work. This is almost a ritualistic experimental work, and, if the operator is not convinced about his challenging task, he will probably never obtain reliable results. There is no doubt about the ubiquity of aluminum and, therefore, handling samples containing aluminum at ppb levels demands very hard work.

From an operational point of view, there are three main sources of extraneous aluminum: air, containers, and reagents. Contamination arising from these sources can be minimized but not totally avoided, and this demands care.

In this chapter, the contamination sources of aluminum and the handling of samples from collection to analyte measurement will be discussed, in order to reach reliable results in the determination of aluminum in the ppb range.

## 3.1
## Contamination Sources

### 3.1.1
### *Air*

Dust particles can be classified according to their size. Those smaller than 0.1 μm behave as a gas and do not deposit. Particles between 0.1 and 1 μm settle, but so slowly that air drafts avoid their deposition; and those bigger than 1 μm deposit at measurable rates. Smoke from fuels and tobacco combustion, viruses and some bacteria are in the first group; flour, clays, paint pigments and bacteria are in the second one; and ash, sand, coal, pollen, spores, hair, and mist are in the last one. In an ordinary atmosphere, 99% of the particles are smaller than 1 μm.

Aluminum exists in dust and its level depends on the environment. Even in areas free from human and industrial emissions, dust particles contain aluminum.

Airborne contaminants existing in a laboratory have a variety of sources, but they are generally a mixture of particles generated within the laboratory itself and those already present in the air that is drawn into the room. The contaminants are normally in gaseous or particulate form, but liquids in form of aerosol can also occur.

There are different ways to remove airborne particles, with different levels of efficiency, which depend on the nature and amount of particles in suspension.

An environment free from atmospheric contamination is generically called a "clean room". However, clean is a relative concept, as are cold or small, and a clean room is not an environment absolutely free from dust particles. As 90% of the airborne particles have a size smaller than 0.5 μm, clean rooms are classified according to the maximal number of particles bigger than 0.5 μm per $ft^3$ they may contain. So, clean environments (rooms, benches or hoods) are classified by numbers, 100, 1000, or 10,000 that mean the number of particles inside them with diameter bigger than 0.5 μm per $ft^3$. A "class 100" clean room, is, therefore, a room with no more than 100 particles with 0.5 μm diameter per $ft^3$ [66].

### 3.1.2
### *Containers*

Laboratory containers are usually made from glass or plastic. Both materials are sources of aluminum. Aluminum oxide is a glass constituent, and plastics can content traces of aluminum due to the use of aluminum compounds as catalysts in many plastic polymerization processes. Other sources of aluminum resulting from sample handling are elastomeric materials. Table 5 presents the aluminum content of different raw materials used in manufacturing labware.

Aluminum is present in all type of glasses, from common glass to those for laboratory uses. The amount of aluminum (and boron as well) added to glass depends on how resistant to chemical attack and temperature changes the glass must be. The higher the aluminum content, the higher the resistance. Aluminum is not supposed to be leached from glass (because it is added just to increase chemical resistance) [72], but, depending on the substance stored in the glass container, aluminum can be picked up from the glass surface. Substances that show an affinity for aluminum ions cannot be stored in glass containers [73, 74], even if the container was very well precleaned. The

**Table 5.** Aluminum content in different raw materials used in manufacturing labware

| Material | | Al present | |
|---|---|---|---|
| | | As content | As impurity |
| Glass [67] | Borosilicate (Pyrex) | 2–7% $Al_2O_3$ | |
| | Common glass | 1.5–6% $Al_2O_3$ | |
| | Common glass amber | 1.5–6% $Al_2O_3$ | |
| | Quartz [75] | | 10–50 μg Al $g^{-1}$ |
| Polyethylene [68] | Low density | | 0.5 to 4.3 μg Al $g^{-1}$ |
| | High density | | 18 to 73 μg Al $g^{-1}$ |
| Polypropylene [68] | | | 53 to 78 μg Al $g^{-1}$ |
| Polytetrafluoroethylene [69] | | 0.2 μg Al $g^{-1}$ | |
| Silicone [70] | Pipe | | 0.03% Al |
| | Rubber | | 0.42% Al |
| Rubber [71] | Halobutyl (gray) | 40% calcined aluminum silicate | |
| | Natural (red) | 29% calcined aluminum silicate | |
| | Black EPDM | – | – |
| | Gray silicone | – | – |

EPDM = ethylenepropylenediene.

aluminum present just on the container wall surface can be washed out by action of the washing solution. However, other aluminum ions can migrate from the inner to the outer surface, and "new" aluminum ions will be able to be leached in an ongoing process. Fig. 12 shows the aluminum leached from glass during one year of contact by the action of complexing agents for aluminum. Even quartz has a small amount of aluminum oxide [75] and should be avoided.

A wide range of plastics with different properties has been used in the construction of laboratory apparatus. In spite of the adsorption ability of the polymers and, therefore, the risk of analyte loss by adsorption on container walls, contamination arising from residual catalysts and additives used in their manufacture is the main problem. Among the most common plastics used in manufacturing laboratory containers are low-density polyethylene (LDPE), high-density polyethylene (HDPE), polypropylene (PP) and polytetrafluoroethylene (PTFE).

LDPE and PTFE are manufactured by processes involving little contact with metals, and should not be a major source of aluminum. On the other hand, HDPE and PP have similar manufacturing processes, based on catalysis, that involve aluminum, among other metals. The contamination risk may therefore be very high. In addition, the manufacture of the container itself can lead to aluminum contamination, considering the use of lubricants, stabilizers and other additives. For example, metal soaps, such as aluminum, magnesium, sodium, and tin, were sanctioned by the FDA to be used as stabilizers in plastic containers for parenteral nutrition solutions [76].

Even polyethylene specially treated for medical implants has aluminum as a contaminant; ultrahigh molecular mass polyethylene, considered of special quality, may present up to 40 ppm Al [77].

Elastomeric materials find use as vial closures, as syringe plungers, and as liquid delivery components. Elastomers, called sometimes simply rubber, are also polymeric materials, and are known by their common names, halobutyl, silicone, urethane, neoprene, polybutadiene, etc. Elastomers themselves

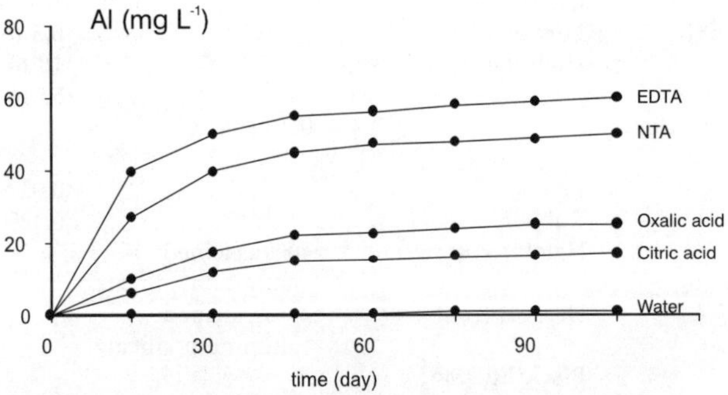


**Fig. 12.** Al extracted from glass containers by action of complexing agents during a year of contact

contain no aluminum. However, rubber manufacturing combines them with other chemicals to produce a vulcanized product having specific physical and chemical properties. Ingredients other than the elastomer are the vulcanizing agents, accelerators, activators, and fillers. Among the materials most frequently used as fillers are hydrated or calcined aluminum silicate. Common red and gray rubber closures contain this as filler whereas black and silicone ones contain carbon black and silica, respectively. Translucent elastomeric materials, called "gum rubber", are formulated without filler and, therefore, present no risk of having aluminum in their composition.

### 3.1.3 *Reagents*

A variety of terms is used to describe the quality of reagents, and this quality is concerned with contaminant impurities. Generally, labels or certificates show the impurity levels, but if an element is not mentioned it does not mean that it is absent. In the recent past, aluminum was not labeled as an impurity (in most "analytical reagent grade" reagents) but currently, it is possible to find, for some products, specially those for pharmaceutical use, specification of aluminum levels. Knowing the contamination level of reagents used in the analysis is important because it can be even higher than the level to be determined in the sample. Table 6 shows the aluminum level found in different reagents of different purity grades. Unfortunately, aluminum is a very common impurity, not only in inorganic but also in organic reagents.

Not only in common reagent substances but also in some for specific uses, aluminum can be present. Exley reported that aluminum was found in commercial ATP (0.34 mmol Al mol $ATP^{-1}$), and how this affected the results of biochemical experiments [78]. Albumin [79–81] and heparin [80–82] are also reported as having high levels of aluminum as contaminant. However, albumin and heparin are special cases of contamination. The aluminum levels found in the products commercialized as solution are much times higher than the levels found in the solid substances. Table 7 shows the aluminum levels in albumin and heparin supplied as solid substances and as solutions from different manufacturers.

Another related source of contamination is formalin, but the use of formalin in tissue conservation can lead to a different problem. Due to the unavailability of medical-grade formalin solution, Meldrun [83] measured the aluminum content of tissues and formalin solution before and 10 days after they had been put in contact. In this case, even presenting aluminum as a contaminant, formalin acted as an aluminum scavenger rather than a contamination source; at the end of the experiment, the metal concentration decreased in the tissue and increased in the formalin solution.

The contamination of substances is difficult to avoid. Purification methods are not easily applied for all substances and purified products are generally expensive.

**Table 6.** Aluminum measured as an impurity in some substances

| Substance | Quality (Merck) | Al (μg Al $g^{-1}$) or (μg $L^{-1}$) | Quality (Brand) | Al (μg Al $g^{-1}$) or (μg $L^{-1}$) |
|---|---|---|---|---|
| HCl | p.a. | 130 ± 3 | p.a. (Merck) | 180 ± 10 |
| $HNO_3$ | p.a. | 110 ± 1 | p.a. (Merck) | 103 ± 4 |
| $HNO_3$ | Suprapur® | 100 ± 1 | | |
| $HNO_3$ | Sub-boiling | 40 ± 1 | | |
| $HNO_3$ | Sub-boiling (double-distilled) | 10 ± 1 | | |
| $H_2SO_4$ | p.a. | 375 ± 3 | | |
| Acetic acid | p.a. | 90 ± 10 | p.a. (Chemco) | 110 ± 10 |
| NaOH | p.a. | 4.13 ± 0.19 | p.a. (Merck) | 5.18 ± 0.20 |
| NaCl | p.a. | 0.39 ± 0.02 | p.a. (Synth) | 0.13 ± 0.01 |
| KCl | p.a. | 0.36 ± 0.01 | USP (Merck) | 0.22 ± 0.02 |
| $CaCl_2$ | p.a. | 2.45 ± 0.24 | p.a. (Merck) | 0.52 ± 0.04 |
| $MgCl_2$ | p.a. | 2.75 ± 0.21 | p.a. (Sigma) | 0.83 ± 0.05 |
| $CH_3COONa$ | p.a. | 0.08 ± 0.01 | USP (Merck) | 0.83 ± 0.03 |
| $NaHCO_3$ | p.a. | 0.16 ± 0.01 | USP (Merck) | 3.21 ± 0.11 |
| $NaH_2PO_4$ | p.a. | 0.61 ± 0.01 | USP (Merck) | 4.92 ± 0.11 |
| $Na_2HPO_4$ | p.a. | 0.30 ± 0.01 | p.a. (Inlab) | 3.91 ± 0.01 |
| $KH_2PO_4$ | p.a. | 0.78 ± 0.03 | USP (Merck) | 3.66 ± 0.12 |
| $K_2HPO_4$ | p.a. | 5.81 ± 0.28 | | |
| Trichloroacetic acid | p.a. | 5.80 ± 2.30 | p.a. (Reagen) | 20.00 ± 4.90 |
| Glucose | p.a. | 0.25 ± 0.10 | p.a. (Sigma) | 0.12 ± 0.08 |
| Citric acid | p.a. | 3.21 ± 0.87 | p.a. (Sigma) | 4.01 ± 1.65 |
| Ascorbic acid | p.a. | 0.15 ± 0.03 | | |
| Chrome azurol S | | | Riedel-deHaën | 2.2 ± 1.4 |

**Table 7.** Aluminum present as an impurity in albumin and heparin in different type of products

| Solid sample | Al content (μg $g^{-1}$) | Solution (Supplier)(n*) | Al content (μg $L^{-1}$) |
|---|---|---|---|
| Albumin | | | |
| Egg (Sigma) | 0.57[b] | 20% (Zanalb) (3) | 94 ± 5[b] |
| Bovine (Merck) | 0.42[b] | 20% (Behring) (3) | 235 ± 12[b] |
| Bovine (Reagen) | 0.32[b] | 5% (Red Cross) (2) | 391 ± 51[a] |
| | | 5% (Travenol) (40) | 163 ± 96[a] |
| | | 5% (Cutter) (6) | 332 ± 92[a] |
| | | 5% (Armour) (8) | 1108 ± 348[a] |
| | | 25% New York Blood (1) | 391 ± 51[a] |
| Heparin (powder) | | Solution 5000 UI/ml | |
| Na salt (bovine) (Sigma) | n.d.[b] | Fujisawa Inc. | 732 ± 23[b] |
| Na salt (pig) (Sigma) | n.d.[b] | Roche | 783 ± 51[b] |
| $NH_4$ salt (pig) (Sigma) | n.d.[b] | Cristália | 72 ± 6[b] |
| | | Eurofarma | 436 ± 40[b] |
| | | Elkins Sinn | 425 ± 27[b] |

* n = n° of samples.
[a] After [79].
[b] After [73].

Some acids can be purified by sub-boiling distillation. Nitric acid, the most important one considering material cleaning and sample decomposition, can be easily purified by this technique. Table 6 shows that the aluminum level in nitric acid can be reduced but not totally eliminated in a single distillation procedure. Non-ionic substances can be purified by using chelating resins, able to retain only the aluminum ions. However, carrying out the purification of a solution without changing its composition is not an easy task. Generally, chelation occurs at a defined pH or a pH range, which may not be that presented by the solution. Consequently, adjusting the pH will change the composition but, by not adjusting it, the removal of the contaminant aluminum ions may be incomplete. Purification of inorganic substances like salts can be done by ion-exchangers, however, unless the cation exchanger has the same counter-ion as the substance to be purified, attempts to remove the aluminum ions can lead to another kind of contamination, by the resin counter-ion.

Exley et al. [61] found aluminum in practically all reagents used in a study of the inhibition of hexokinase activity by this element. The way to overcome the problem was cleaning the solutions using an aminophosphonate chelation resin. The procedure reduced the contamination of ATP and NADP to approximately 5% and 10% of their initial values, respectively, but the resin was ineffective in removing aluminum from magnesium acetate or the enzyme glucose 6-phosphate dehydrogenase. Probably the conditions were not favorable for the resin to pick up the aluminum ions from these solutions. It is important to remember that, if there is an affinity between aluminum and the species in solution, a competition between this species and the resin will take place.

Sanz-Medel and coworkers [84] developed a system to avoid aluminum contamination in a chromatographic system, used for studying aluminum bound to serum proteins, by introducing an on-line Al-chelating scavenger pre-column.

In order to lower the blank in a chromatographic procedure for the determination of trace amounts of Al citrate, Bantan et al. [85] purified the eluent and the chromatographic column itself. The eluent, a solution of $NH_4NO_3$, was purified in an ion-exchange batch procedure, with Chelex 100, Na form, for 24 h, followed by a column cleaning step. The pre-purified eluent passed through a silica-based RP-18 column to remove the remaining aluminum traces. The column was cleaned with a injection of 500 μL of 5 mol $L^{-1}$ citric acid at least 5 times, prior to applying the sample, until the Al citrate concentration was lowered to less than 1 μg $L^{-1}$.

Rowatt and Williams [86] used an ion-exchange membrane (Bio-Rex) to remove polyvalent cations from phosvitin, a phosphorylated protein, before studying its interaction with aluminum in the presence of silicic acid.

Water is the main reagent in any analysis. Neither distilled water nor deionized water are pure enough to be used in trace analysis. Conventional distillation reduces the contamination, but the level of purity which can be attained depends on the still design and operation. Up to 10 μg $L^{-1}$ Al can be found in distilled water [87]. The use of ion-exchangers is an effective

**Table 8.** Other contamination sources in trace aluminum determination

| Aim | Contamination source | Contamination control | Reference |
|---|---|---|---|
| Al-binding proteins | Ultrafiltration membranes | Washing with: 1 mol $L^{-1}$ sodium citrate + water + 0.1 mol/L NaOH + water and 500 mg $L^{-1}$ DFO | [88] |
| Al distribution in serum | Ultrafiltration system<br>Gel filtration column | UF system: washing with 10% $HNO_3$, 24 h<br>membranes: twice 0.1 mol $L^{-1}$ NaOH<br>gel filtration column: 30 mol $L^{-1}$ $Na_2CO_3$, 48 h | [89] |
| Al determination in serum | Chromatographic system | Stainless-steel free chromatographic system<br>cleaning column: 1.5 mM DFO, 12 h | [90, 91] |
| Speciation Al complexes in serum | Chromatographic system | Cleaning eluent ($NH_4NO_3$): ion-exchange batch procedure<br>Chelex 100, Na-form, 24 h<br>Column: injection of 5 mol $L^{-1}$ citric acid, 5 times | [92] |
| Al determination in serum | Ultrafiltration membranes | Washing with 5 mL portions of 5% $HNO_3$ and water until negative test for Al | [93] |

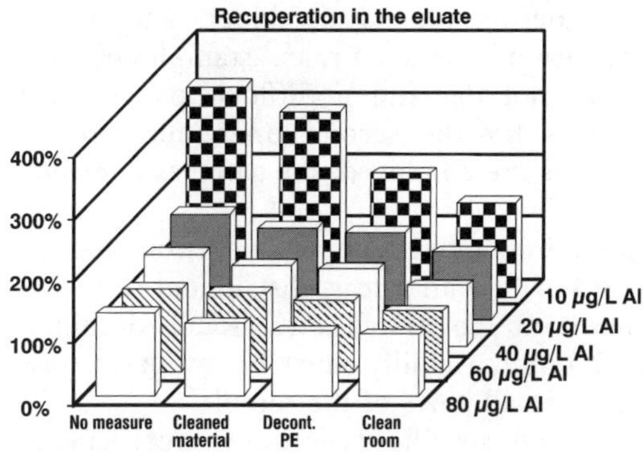


**Fig. 13.** Aluminum recovered in samples containing from 10 to 80 $\mu L^{-1}$ Al after adopting stepwise measures to avoid contamination from containers, reagents, other materials and dust in air. Data are from Garcia et al. (1997) [94]

means of removing charged contaminants, but regarding aluminum, the existence of charged species is highly pH-dependent. Therefore, aluminum can exist as a neutral species unable to be retained by the exchanger. Because each technology removes a specific species or particle, neither can be relied upon to remove all contaminants to the levels required for this application. A purification system using a combination of technologies (filtration, deionization, membrane filtration, etc.) may achieve the necessary final water quality.

Sources of aluminum other than reagents were related by some authors and are presented in Table 8. Ultrafiltration membranes and package materials for chromatographic columns are listed, although their use without additional care to lessen contamination, could lead to false results [88–93].

Fig. 13 shows an example of how important it is to avoid contamination, and its influence on the final result of the analysis. The analytical procedure was a pre-concentration of aluminum by adsorption of its complex with an organic reagent (chrome azurol S) onto a polymeric material (polyethylene powder) packed into a column. Standard solutions were prepared and the procedure carried out after adopting stepwise precautions to avoid contamination. It can be seen that, only after adopting all steps, was the contamination controlled. The graph also shows that the lower the aluminum content in the sample the higher is the contribution of the contamination sources.

## 3.2 Sample Collection and Preparation

Sample collection and preparation covers *in vitro* and *in vivo* studies, and the later involves humans or model animals. In this field the samples can be materials with very different characteristics and features. From an analytical point of view, samples could be divided into two categories: liquid samples and solid samples. Considering biologicals, blood and urine are in the first category and tissues, from bone to hair, are examples of the second. Generally, the first group does not demand a difficult pretreatment before analyte measurement, whereas for the second one, unless techniques that allow analyzing solid samples are employed, the samples demand a more laborious pretreatment.

Another consideration about sample collection and preparation is the interest in knowing how the analyte is distributed in the sample. If only the total amount is important, the sample can be totally decomposed. On the other hand, if the analyte binding ability must be evaluated, pretreatment should keep analyte integrity in all parts of the sample.

Because there are no specific procedures but general concerns, some specific procedures for sample collection and preparation will be commented on in this chapter.

Table 9 summarizes procedures of blood collection for aluminum determination. Steps to avoid contamination, the analytical technique, and the aluminum levels found in the serum of healthy individuals are listed. It can be seen that, as a general procedure, plasticware was used, all materials that entered in contact with the sample were acid-washed, usually by nitric acid, and the analyses were carried out inside clean rooms.

Considering more complex samples or samples that demand a more laborious pretreatment, a more laborious contamination control should be adopted. Slavin and coworkers [104], in order to determine aluminum in bone, wore powderless gloves and used tantalum knives and scrapers (made from ultra-pure tantalum, relatively free from aluminum) to remove adhering tissue from bone. They also used a diamond-disc saw to cut small segments of the sample for analysis, besides using only plasticware and sub-boiling distilled $HNO_3$.

Nordahl and coworkers [105] used direct solid sampling to determine aluminum in human biopsy and necropsy specimens by ETAAS. They commented that the main difficulty in the determination of aluminum in solid samples was neither lack of instrumental sensitivity, nor interferences, but rather the prevention and control of contamination. The authors had to modify the cup in which the sample was introduced into the graphite tube in order to reduce contamination, besides all the other steps, such as washing all materials with 0.6% EDTA solution, covering the balance and the working area with Parafilm, working under class 100 conditions, and executing an instrumental tube clean-up between sample runs. Fig. 14 shows the standard and modified cup for solid sampling.

Pierson [106] washed all materials with 0.1 mol $L^{-1}$ EDTA solution and used Teflon-coated dissection tools and tubes to excise and store neuronal tissues.

Domingo et al. [107, 108] used ultrapure nitric acid for sample digestion, performed all specimen manipulation in a laminar flow hood, carried out readings of a blank (reagents only) for each group of samples and, to minimize matrix effects, prepared the standard curves for aluminum determination by the standard addition method.

**Table 9.** Care in sample collection for measurement of aluminum levels in the blood of healthy subjects

| Technique | Blood collection | Serum collection | Clean room | Washing solution | Extra cautions | Al μg $L^{-1}$ | Reference |
|---|---|---|---|---|---|---|---|
| ETAAS | Acid-washed glass | PE | Class 100 | 5% $HNO_3$, 12 h | Yes | 0.54–3.51 | [95] |
| ETAAS | Red-top Vacutainer | PP | – | – | – | 3.6 | [96] |
| HPLC | – | – | – | 1 mol $L^{-1}$ $HNO_3$, overnight | – | 2 | [97] |
| ETAAS | Plastic tubes | Plastic tubes | – | – | – | 4.40 ± 2.84 | [98] |
| HPLC | – | Acid-washed tube | – | – | – | 5.5 ± 3.7 | [99] |
| CZE | – | PP | – | – | – | 14 | [100] |
| HPLC | PP | Eppendorf vial | – | – | – | 2.2–15.7 | [101] |
| ETAAS | Monoject LH 10 | Acid-washed tube | Class 10 | Ultrapure acids | Yes | 4.10–11.39 | [102] |
| ETAAS | Acid-washed PP | Acid-washed tube | Class 10 | Ultrapure acids | Yes | 0.81–1.19 | [102] |
| ETAAS | Al-free Vacutainer | Teflon | Class 10000 | 10% $HNO_3$, 24 h | – | 3.5 ± 0.5 | [85] |
| ETAAS | Metal-free PP | PP | Class 100 | 0.1% alkaline phosphate free detergent | Yes | 3 ± 1 | [103] |

– = not mentioned.

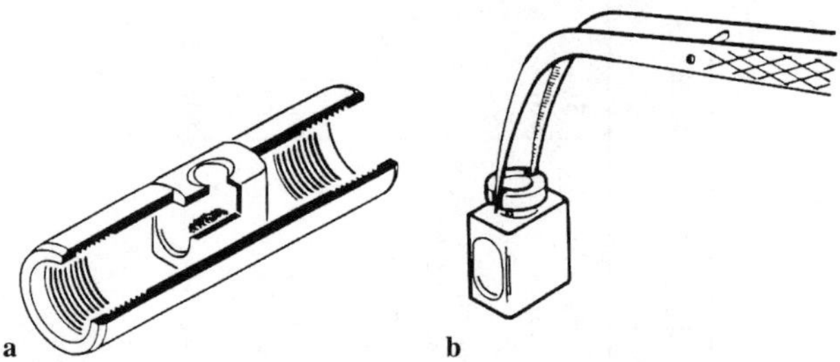

**Fig. 14.** Standard (**a**) and modified (**b**) cup for solid sampling for Al measurement in biological tissues. Data are from Nordahl et al. (1990) [105]

As a general rule, samples should be collected in decontaminated flasks and even those considered metal-free should be pre-washed. Glass recipients should be avoided. A good procedure for cleaning plastic containers is keeping them at least 24 h in a 10% (v/v) ethanolic nitric acid solution. Aqueous solutions do not leach out the aluminum well, because water does not wet plastic surfaces; better contact occurs when an alcoholic solution is used. It can be seen that all alcoholic solutions were more efficient than the aqueous solutions. Table 10 shows the aluminum extracted from polyethylene by the action of some washing solutions. Just before use, the containers should be abundantly washed with ultrapure water. The best option is to use the flasks just after rinsing them, however if they must be dried, they should not be placed in an oven, even with the open side on tissue paper (paper contains aluminum). The best way is let them dry under a laminar flow. The proper heat and air movement inside the hood will help to rapidly dry the flasks.

Fig. 15 shows the aluminum extracted from PE powder and glass powder by action of aqueous and ethanolic $HNO_3$ solutions during 7 days of contact. It can be seen that 24 h is enough to withdraw the aluminum from the PE surface when a ethanolic solution of $HNO_3$ is used. The aqueous solution acts slower because PE and the solution do not have good contact. With glass, the contact is effective but the extraction process did not end even after 7 days.

**Table 10.** Aluminum extracted from polyethylene powder by action of different washing solutions by shaking 0.1 g PE in 10 mL solution for 24 h (blank discounted)

| Extraction solution | Extraction rate ($\mu g\ g^{-1}$) |
|---|---|
| Water | 0.3 |
| Ethanol | 2.2 |
| 0.1 mol $L^{-1}$ EDTA | 4.1 |
| 10% (v/v) aqueous $HNO_3$ | 2.6 |
| 0.2% (v/v) ethanolic $HNO_3$ | 7.0 |
| 10% (v/v) $HNO_3$ ethanol:water (1:1) | 7.4 |
| 10% (v/v) ethanolic $HNO_3$ | 9.7 |

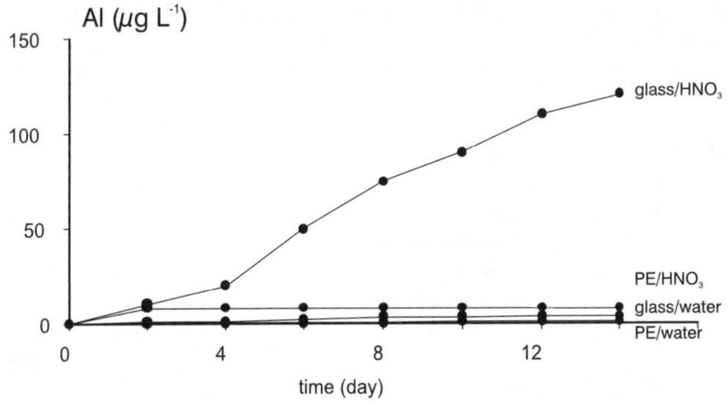


**Fig. 15.** Al leached from glass and PE by action of aqueous and ethanolic solutions of $HNO_3$ during 7 days contact

Collection of biological samples as tissues or organs should be carried out carefully, mainly when small animals are the study object. Gloves should of course be used, but they must be washed to remove the talc before handling the analysis apparatus. The animals should also be washed, with a detergent, followed by tap water and then purified water. The animals can be dried with tissues or cotton towels, although both are aluminum sources. A guillotine can be used for organ collection, but not for blood. For blood collection two ways can be considered. When the animal is big enough to collect blood from a vein, special commercially available tubes for metal-free blood collection should be avoided, they are made from glass which contains aluminum. These tubes are closed with special rubber tips that contain more aluminum than the tube; therefore they should not get in contact with the sample. Table 11 shows the aluminum content, as constituent or impurity, in different materials used for blood sample collection or handling. Needles cannot be avoided, even though they are possible aluminum sources. Those made of high-quality, surgical grade, stainless steel may have to be used.

Small animals, for which parenteral blood collection is not possible, should be killed with tools made of plastic or ceramic. There are special tools (scissors) made from zirconium, ceramic or titanium ceramic for hair cutting that can also be used for trace analysis.

All tools should be washed with alcoholic nitric acid, water and be well dried before use. They should be placed under laminar flow at least half hour before starting sample collection.

Not only are clean reagents important for a successful analysis, reducing their amount can also be a good option. Slavin et al. [104] treated bone samples with a mixture of methanol/diethyl ether (1 + 1) overnight, in order to defat them. This procedure may reduce the amount of acid needed for sample digestion, because lipids are, among biological constituents, the more resistant to be decomposed.

Constantini et al. [109] dried and ashed the tissue samples before digestion and removed fat from the tibiae with a mixture of chloroform and methanol

**Table 11.** Aluminum present in different materials used in blood collection and leached after 7 days contact with water

| Material | Object | Al present (% m/m) | Al leached (μg $L^{-1}$) |
|---|---|---|---|
| Glass | Glass flask clear | 0.59 | 16 ± 2 |
| | Glass flask dark | 0.52 | 10 ± 2 |
| | Vacutainer (Royal blue tip) (Metal trace analysis) | 0.73 | 12 ± 1 |
| | Monoject (Lavender tip) | 1.22 | 19 ± 3 |
| | Monoject (No Additive) | 1.31 | 11 ± 2 |
| | Venoject (Silicone coated) | 1.45 | 51 ± 6 |
| Plastic | Polyethylene flask (LDPE) | n.a. | <1 |
| | Syringe 10 mL | n.a. | 2 ± 1 |
| | Microtainer 365962 | n.a. | 3 ± 1 |
| | Monovette (Metal trace analysis) | n.a. | 2 ± 2 |
| | Eppendorf vial (1.5 mL) | n.a. | 4 ± 4 |
| | AAS sample cup | n.a. | 3 ± 4 |
| | Pipette tip yellow (100 μL) | 0.01 | 3 ± 1 |
| | Pipette tip (1000 μL) | 0.01 | 4 ± 1 |
| Rubber | Common gray closure | 3.9 | 5 ± 2 |
| | Common red closure | 5.3 | 9 ± 3 |
| | Royal blue tip (Vacutainer) (Metal trace analysis) | 4.6 | 3 ± 2 |

n.a. = not analyzed.

(1 + 1). They also carried out the sample preparation under dust-free conditions, used stainless steel surgical instruments and checked all equipment for specimen collection before use.

Bohrer et al. [93] used microwave heating as an adjuvant to precipitate serum proteins prior to ETAAS aluminum determination. With this procedure, the concentration of TCA necessary to precipitate more than 99% of the proteins could be reduced to 10% of the amount needed without such microwave pretreatment.

An example of the influence of extraneous aluminum in an *in vivo* study with animals can be seen in Fig. 16. The graph shows the results of aluminum determination in brain and liver of mice in a study of the effect of aluminum sulfate on delta-aminolevulinate dehydratase from these organs [110], before and after measures to avoid aluminum contamination had been adopted. In a first attempt to check the way to kill the animals and how to digest the samples, the animals were sacrificed with no measures taken to avoid contamination. The results from this experiment, when compared with those after controlling extraneous aluminum as much as possible, show an aluminum level at least 10 times higher. The results also show that, due to the extent and to the randomness of the distribution of the contamination, there are no significant differences between controls and Al-treated animals, when no attempts to avoid contamination are adopted.

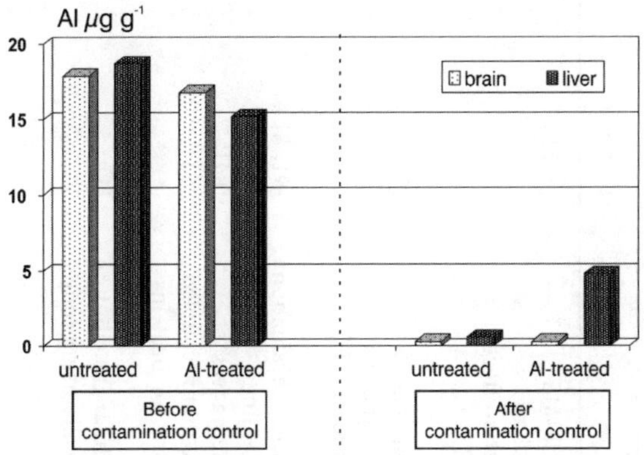


**Fig. 16.** Al distribution in brain and liver of treated and untreated mice, with and without measures to avoid extraneous Al contamination

## 3.3 Techniques for Aluminum Determination

As mentioned for sample collection, the choice of the analytical technique is conditioned by the answer to the question: What information about the analyte is wanted?

Nowadays, not only "how much" but also "where" and "in what fashion" is the analyte in the sample are questions to be answered.

The possibility of combining separation and detection techniques gives origin to the so-called "hyphenated-techniques", analytical systems able to answer these questions. Well described approaches on the subject are in the special issue of Spectrochimica Acta devoted to Speciation [111].

An overview on advances in instrumental methods for the measurement and speciation of trace metals is given by Savory [112]. He comments about the analytical techniques that help understanding the role of trace metals in biology and medicine.

Analytical methods for aluminum can be divided into those that give information about its localization, and those that inform about its quantity. In the first group are nuclear techniques, involving the Al-26 tracer and the Al-27 nuclear magnetic resonance (NMR). In the second group is the well-established technique for aluminum quantification, ETAAS, and a new generation of chromatographic separations coupled with spectrometric detection that allow compartmentalized quantification, the hyphenated systems.

Electrothermal atomic absorption spectrometry (ETAAS) was in the past, and still is in the present, the most useful technique for total aluminum determination in biological samples. The technique is subject to contamination from dust, but by keeping the instrument under laminar flow conditions or setting the sample in the autosampler just before pipetting, contamination may be avoided. In recent years, chromatographic techniques have also been

**Table 12.** Analytical methods for aluminum investigation in biological and biological related matrices

| Technique | Matrix | Aim | Sample handling/comments | Reference |
|---|---|---|---|---|
| ICP-MS and ETAAS | Serum | Al determination | Comparison of ETAAS and two ICP-MS arrangements, quadrupole-based and double-focusing, in terms of matrix-effect, spectral interferences and overall analytical performance | [115] |
| HPLC-ETAAS | Cerebrospinal fluid | Al speciation in patients treated with DFO | Evaluation of Al binding species in acutely Al-intoxicated dialysis patients by analysis of cerebrospinal fluid before and after DFO administration | [116] |
| HPLC | Serum | Al speciation after DFO administration | By means of UF and HPLC techniques evaluation of Al-species other than alumioxamine | [91, 8] |
| FPLC-ICP-MS | Serum | Al bound to proteins | Separation of human serum proteins was achieved on an anion-exchange column coupled with a spectrophotometric detector for proteins and a ICP-MS for Al | [84] |
| ETAAS | Neuronal and astroglial cells | Preferential Al accumulation | Analysis of the uptake and subcellular distribution of Al in neuronal and astroglial cells using four different Al salts | [1] |
| ETAAS | Water | Relation between long-term exposure to Al and AD development | Speciation of Al in drinking water in 54 municipalities for a year and evaluation of 70-year-old or more residents status related to clinical AD diagnosis | [117] |
| Potentiometry | Aqueous solutions | Al binding ability | Examination of Al bioavailability through speciation of Al complex formation with ligands at physiological conditions | [24, 30, 118–121] |

| | | | | |
|---|---|---|---|---|
| Potentiometry and NMR | Aqueous solutions | Interaction of Al with glucose 6-phosphate | Evaluation of the species formed in the system Al-glucose 6-phosphate at different pH | [29] |
| EELS and ESI | Biological tissues | Al determination | Description of the application of the techniques to Al determination in biological tissue | [122] |
| AMS-Al-26 | Serum, urine, bone, liver, spleen | Intestinal absorption of Al in rats with renal failure | Trace quantities Al-26 were given in single oral dose to the animals, and by analysis of the fluids and tissues the intestinal absorption and compartmentalization of Al was evaluated | [123] |
| AMS-Al-26 | Blood, urine | Technical parameters evaluation | Study of interference and materials to be used as source of Al-26. Evaluation of limit of detection, reproducibility and precision of the technique | [124] |
| AMS-Al-26 | Blood | Gastrointestinal Al absorption in patients with Alzheimer's disease | Volunteer patients ingested a drink containing Al-26 and the blood, collected before and after ingestion, was assayed by AMS to determine the fraction of Al absorbed across the gastrointestinal tract | [125] |
| AMS-Al-26 | Rat brain | Distribution of Al into and out of the brain | Studies of extent, rate and possible mechanism(s) by which Al enters and is removed from the brain through introduction of Al-transferrin and Al-citrate dose into systemic circulation. Evaluation of Al-26 concentrations in brain and in extracellular blood | [126] |
| AMS-Al-26 | Brain | Separation of Al and phosphorus | Removal of phosphate from acid digested brain samples in order to heave a clean $Al_2O_3$ sample to AMS analysis | [127] |

employed for total aluminum determination in biological samples. Comparatively, in reviews from 10 or 20 years ago, the only described methodology for aluminum determination in blood was the AAS [113, 114]. In Table 9, a collection of methods for aluminum blood analysis in the last 10 years presents 4 chromatographic techniques in a total of 11 methods described. On-line chromatographic techniques show the advantage, compared to ETAAS, of a clean environment, i.e., if the system does not have metallic parts and has been cleaned, it remains free of external contamination. On the other hand, prior to the injection of the sample it is necessary to remove high molecular mass compounds to prevent clogging of the columns. In ETAAS the removal of this kind of sample component is not essential, and an adequate ashing temperature overcomes this problem.

As the possible combinations of samples and analysis techniques are very large, Table 12 presents an overview of these combinations related in the literature.

## 4 Concluding Remarks

In this review we intended to join information about some chemical, biological, toxicological, and analytical aspects of the aluminum. The aim of this chapter is not to give new information for specialists, but to help inexperienced researchers who work with this element in biological systems. Many basic analytical aspects involved in the aluminum determination in biological matrices are not available in the current literature. Thus, we hope to have reached our goal and really contributed to the understanding of the aspects cited above. In a chapter of this nature, there are bound to be errors, for which we wait for comments and suggestions from readers to improve our review.

**Acknowledgement.** We have benefited from the goodwill and help of a number of colleagues, especially to João Batista Teixeira da Rocha, Paulo Cícero do Nascimento, Carol Collins and Daniela Balz. We should like to acknowledge the following for permission to reproduce their illustrations. Secondly, all those publishers who have granted us formal approval: Sociedade Brasileira de Química: Fig. 13, Elsevier Science Inc.: Fig. 1, Fig. 2, Fig. 3, Fig. 4, Fig. 5, Fig. 6, Fig. 7, Fig. 8, Fig. 9, Fig. 11, Table 1, Table 2, Table 3, Table 4. Humana Press: Fig. 10, Springer-Verlag: Fig. 14.

## 5 References

1. Lévesque L, Mizzen CA, McLachlan R, Fraser PE (2000) Brain Res 877: 191
2. Williams RJP (1999) J Inorg Chem 76: 81
3. Chadwick DJ, Whelan J (eds) (1992) Aluminum in Biology and Medicine, Ciba Foundation Symposium 169, Wiley, Chichester
4. Williams RJP (1996) Coord Chem Rev 149: 1
5. Atwood DA, Yearwood BC (2000) J Org Chem 600: 186

6. Whitehead MK, Farrar G, Christie GL, Blair JÁ, Thompson RPH (1997) Am J Clin Nutr 65: 1446
7. Berthon G (1996) Coord Chem Rev 149: 241
8. Corain B, Nicolini M, Zatta P (1992) Coord Chem Rev 112: 33
9. Kiss T (1995) Arch Geront Ger 21: 99
10. Yokel RA, Rhineheimer SS, Brauer RD, Sharma P, Elmore D, McNamara PJ (2001) Toxicol 161: 93
11. Kochian LV (1995) Annu Rev Plant Physiol Plant Mol Biol 46: 237
12. Pan JW, Zhu MY, Chen H (2001) Environ Exp Bot 46: 71
13. Severi A (1997) Environ Exp Bot 37: 53
14. Rai LC, Husaini Y, Mallick N (1998) Aquat Toxicol 42: 67
15. Yokel RA, McNamara PJ (2001) Pharmacol Toxicol 88: 159
16. Harris WR, Berthon G, Day JP, Exley C, Flaten TP, Forbes WF, Kiss T, Orvig C, Zatta PF (1996) J Toxicol And Environ Health 48: 543
17. Kiss T, Zatta P, Corain B (1996) Coord Chem Rev 149: 329
18. Nelson DJ (1996) Coord Chem Rev 149: 95
19. Froment DPH, Molitoris BA, Buddington B, Miller N, Alfrey AC (1989) Kidney Int 36: 978
20. Slanina P, Frech, W, Eskon L, Loof L, Slovach S, Cedergen A (1986) Clin Chem 32: 539
21. Wang X, Nelson DJ, Trindle C, Martin RB (1997) J Inorg Biochem 59: 7
22. Weiner ML, Salminen WF, Larson PR, Barter RA, Kranetz JL, Simon GS (2001) Food Chem Toxicol 39: 759
23. Öhman LO, Martin RB (1994) Clin Chem 40: 598
24. Daydé S, Fillela M, Berthon G (1990) J Inorg Biochem 38: 241
25. Dyrssen D, Haraldsson C, Nyberg E, Wedborg M (1987) J Inorg Biochem 29: 67
26. Huang H, Liu M, Mao XA (1998) Spectrochim Acta Part A 54: 999
27. Waldron-Edward D, Chan D, Skoryna SC (1971) Can Med Assoc J 105: 1297
28. Jackson GE, Voyi K (1987) Polyhedron 6: 2095
29. Champmartin D, Rubini, P, Lakatos A, Kiss T (2001) J Inorg Biochem 84: 13
30. Venturini-Soriano M, Berthon G (2001) J Inorg Biochem 85: 143
31. Gandolfi L, Stella MP, Zambenedetti P, Zatta P (1998) Biochim Biophys Acta 1406: 315
32. Julka D, Gill K (1996) Biochim Biophys Acta 1315: 47
33. Wyse AT, Bavaresco CS, Bandinelli C, Streck EL, Franzon, R, Dutra-Filho CS, Wagner M (2001) Neurochem Res 26: 515
34. Zatta P, Zambenedetti P, Pizziuti A, Perazzolo, M (1995) Neurosci Lett 197: 65
35. Lai JCK, Guest JF, Leung TKC, Lim L, Davison, AN (1980) Biochem Pharmacol 29: 141
36. Martin RB (1986) Clin Chem 32: 1797
37. Macdonald TL, Humpreys WG, Martin RB (1987) Science 236: 183
38. Swegert CV, Dave KR, Katyare SS (1999) Mechanisms of Ageing and Development 112: 27
39. Zatta P, Taylor A, Zambenedetti P, Milacic R, dell'Antone P (2000) Life Sci 66: 2261
40. Jones DL, Kochian LV (1997) FEBS 400: 51
41. Ma JF, Ryan PR, Delhaize, E (2001) Trends Plant Sci 6: 1385
42. Hamilton CA, Good AG, Taylor GJ (2001) Plant Physiol 125: 2068
43. Taylor GJ, Basu A, Basu U, Slaski JJ, Zhang G, Good A (1997) Plant Physiol 114: 363
44. Tu S-I, Brouilette JN (1987) Phytochemistry 26: 65
45. Korchazhkina O, Wright G, Exley C (1999) J Inorg Biochem 76: 121
46. Matzel LD, Rogers RF, Talk AC (1996) Neuroscience 74: 1175
47. Miller JL, Hubbard CM, Litman BJ, Macdonald TL (1989) J Biol Chem 264: 243
48. MacDonald TL, Martin RB (1988) TIBs 13: 15
49. Landino LM, Macdonald TL (1997) J Inorg Biochem 66: 99
50. Chabre M (1989) Biochem J 258: 931
51. Coleman DE, Berghius AM, Lee E, Linder ME, Gilman AG, Sprang SR (1994) Science 265: 1405
52. Shafer TJ, Mundy WR (1995) Gen Pharmacol 26:889

53. Fraser SP, Barrett P, Moon C, Morgan PJ, Djamgoz MBA (1994) Molec Neuropharmacol 3: 175
54. Moon C, Fraser SP, Djamboz MBA (1997) Cell Signal 9: 497
55. Shafer TJ, Nostrandt AC, Tilson HÁ, Mundy WR (1994) Biochem Pharmac 47: 1417
56. Li L, Fleming N (1999) FEBS 461: 1
57. Nostrandt AC, Shafer TJ, Mundy WR, Padilla S (1994) The Toxicologist 14: 351
58. Zimmermann H, Braun N, Kegel B, Heine P (1998) Neurochem Int 32: 421
59. Schetinger MRC, Wyse ATS, Da Silva LB, Barcellos CK, Dias RD, Sarkis JJF (1995) Biol Trace Elem Res 50: 209
60. Burnstock G (1997) Neuropharmacol 36: 1127
61. Exley C, Price NC, Birchall JD (1994) J Inorg Chem 54: 297
62. Stewart DC, Copeland L (1993) Phytochemistry 32: 27
63. Cochran M, Elliott DC, Brennan P, Chowtur V (1990) Clin Chem Acta 194: 167
64. Exley C, Birchall JD (1992) Polyhedron 11: 1901
65. Calabrese EJ, Baldwin LA (2001) Trends Pharmacol Sci 22: 285
66. Moody JR (1982) Anal Chem 54: 1358 A
67. Scholze H (1988) Glass: Natur, Struktur und Eigenschaften. Springer, Berlin Heidelberg, New York
68. Fujiwara S, Narasaki H (1961) Japan Analyst 10: 1268
69. Moody JR, Lindstrom RM (1977) Anal Chem 49: 2264
70. Fujiwara S, Narasaki H (1964) Anal Chem 36: 206
71. Smith EJ, Nash RJ (1986) Elastomeric Closures for Parenterals. In: Pharmaceutical dosage forms: Parenteral Medications, vol 2. Marcel Dekker, New York, p 155
72. Bacon FR (1986) Glass Containers for Parenterals. In: Pharmaceutical dosage forms: Parenteral Medications, vol 2. Marcel Dekker, New York, p 85
73. Bohrer D, do Nascimento PC, Binotto R, Pomblum SC (2001) Trace Elem Med Biol 15: 95
74. Bohrer D, do Nascimento PC, Binotto R, Carlesso R (2001) Trace Elem Med Biol 15: 103
75. Tschöpel P, Kotz L, Schulz W, Veber M, Tölg G (1980) Fresenius J Anal Chem 302: 1
76. Solomon DD, Jurgens RW, Wong KM (1986) Plastic Containers for Parenteral. In: Pharmaceutical dosage forms: Parenteral Medications, vol 2. Marcel Dekker, New York, p 111
77. Ullman's Encyclopedia of Industrial Chemistry (1982) 5th edition, vol 19, Wiley-VCH, Weinheim
78. Exley C, Birchall JD (1996) J Inorg Biochem 63: 241
79. Milliner DS, Shinaberger JH, Shuman P, Coburn J (1985) New Engl J Med 312: 165
80. Klein GL (1991) Nutr Rev 49: 74
81. Gammelgaard B, Sandberg E (1989) Trace Elem Electrolytes Health Dis 3: 39
82. Baydar T, Aydin A, Duru S, Isimer A, Sahin G (1997) Clin Toxicol 35: 277
83. Meldrum RD (2001) J Biomed Mater Res 57: 59
84. Cabezuelo ABS, Bayón MM, González EB, Alonso JIG, Sanz-Medel A (1998) Analyst 123: 865
85. Bantan T, Milaèiè R, Philar B (1998) Talanta 47: 929
86. Rowatt E, Williams RJP (1996) Coord Chem Rev 149: 167
87. Hughes RC, Mürau PC, Gundersen G (1971) Anal Chem 43: 691
88. Fernández-Martín JL, Menéndez-Fraga P, Canteros MA, Díaz-López JB, Cannata-Andía JB (1994) Clin Chim Acta 230: 137
89. Wróbel K, González EB, Sanz-Medel A (1993) J Anal At Spectr 8: 915
90. Menéndez-Fraga P, Fernández-Martín JL, González EB, Cannata-Andía JB (1998) Clin Chem 44: 1262
91. Canteros-Picotto A, Fernández-Martín JL, Cannata-Andía JB (2000) Am J Kidney Dis 36: 969
92. Polak TB, Milaèiè R, Mitroviè B, Benedik M (2001) J Pharm Biomed Anal 26: 189
93. Bohrer D, do Nascimento PC, Pomblum SG (1998) J Anal Atom Spectr 13: 635

94. Bohrer do Nascimento D, Garcia SC, Gioda A (1997) Química Nova 20: 407
95. Valkonen S, Aitio A (1997) Sci Total Environ 199: 103
96. Gitelman HJ, Alderman FR, Kurs-Lasky M, Rockette HE (1995) Ann Occup Hyg 39: 181
97. Sato M, Yoshimura H, Shimmura T, Obi H, Hatakeyama S, Kaneko E, Hoshino H, Yotsuyanagi (1997) J Chrom A 789: 361
98. van Rensburg SJ, Potocnik FCV, Kiss T, Hugo F, van Zijl, Mansvelt E, Carstens ME, Theodorou P, Hurly PR, Emsley RA, Taljaard JJF (2001) Brain Res Bull 55: 319
99. Mathieu C, Mathieu H (1995) Clin Chem 41: 625
100. Takatsu A, Eyama S, Uchiumi A (1995) Chromatographia 40: 125
101. Wu J, Zhou CY, Chi H, Wong MK, Lee HK, Ong HY, Ong CN (1995) J Chrom B 663: 247
102. Hewitt CD, Winborne K, Margrey D, Nicholson JRP, Savory MG, Savory J, Wills MR (1990) Clin Chem 36: 1466
103. Granadillo VA, Tahán JE, Salgado O, Elejalde LE, Rodríguez-Iturbe B, Romero GB, Romero RA (1995) Clin Chim Acta 233: 47
104. Tang S, Parsons PJ, Slavin W (1996) Analyst 121: 195
105. Nordahl K, Radziuk B, Thomassen Y, Weberg R (1990) Fresenius J Anal Chem 337: 310
106. Pierson KB, Evenson MA (1986) Anal Chem 58: 1744
107. Domingo JL, Gomez M, Llobet JM, Corbella, J (1991) Kidney Int 39: 598
108. Gómez, M, Esparza JL, Domingo, JL, Singh PK, Jones MM (1999) Toxicology 137: 161
109. Constantini, S, Giordano R, Ioppolo A, Mantovani A, Ballanti P, Mocetti P, Bonucci E (1989) Pharm & Toxicol 64: 47
110. Schetinger MRC, Bonan CD, Morsch VM, Bohrer D, Valentim LM, Rodrigues SR (1999) Braz J Med Biol Res 32: 761
111. Spectrochimica Acta Part B (1998) 53
112. Savory J, Herman MM (1999) Ann Clin Lab Sci 29: 118
113. Versieck J, Cornelis R (1980) Anal Chim Acta 116: 217
114. Schaller KH, Valentin H (1984) Aluminum. In: Alessio L, Berlin A, Boni M, Roi R (eds), Biological indicators for the assessment of human exposure to industrial chemicals. Commission of the European Communities, p 21
115. Muniz CS, Marchante-Gayon JM, Alonso JIG, Sanz-Medel A (1998) J Anal At Spectr 13: 283
116. vanLandeghem GF, DHaese PC, Lamberts LV Barata JD, DeBroe ME (1997) Nephrol Dialysis Transplant 12: 1692
117. Gauthier E, Fortier I, Courchesne F, Pepin P, Mortimer J, Gauvreau D (2000) Environ Res 84: 234
118. Desroches S, Daydé S, Berthon G (2000) J Inorg Biochem 81: 301
119. Desroches S, Biron F, Berthon G (1999) J Inorg Biochem 75: 27
120. Venturini-Soriano M, Berthon G (1998) J Inorg Biochem 71: 135
121. Venturini M, Berthon G (1989) J Inorg Biochem 37: 69
122. Xie XZ, Yokel RA, Markesbery WR (1994) Biol Tr Elem Res 40: 39
123. Ittel TH, Steinhausen C, Kislinger G, Kinzel S, Nolte E, Sieberth HG (1997) Nephrol Dialysis Transplant 12: 1369
124. King SJ, Oldham C, Popplewell JF, Carling RS, Day JP, Fifeld LK, Cresswell RG, Liu KX, diTata ML (1997) Analyst 122: 1049
125. Moore PB, Day JP, Taylor GA, Ferrier IN, Fifeld LK, Edwardson JA (2000) Dementia Geriatr Cognit Disord 11: 66
126. Yokel RA, Allen DD, Ackley DC (1999) J Inorg Biochem 76: 127
127. Brauer RD, Robertson JD, Sharma P, Yokel RA (1999) Nucl Instr Meth Phys Res 152: 129

# Aluminofluoride Complexes in the Etiology of Alzheimer's Disease

Anna Strunecká[1], Jiri Patočka[2]

[1] Charles University Prague, Faculty of Science, Department of Physiology and Developmental Physiology, Vinicna 7, 128 00 Prague 2, Czech Republic
*E-mail: strun@natur.cuni.cz*

[2] Military Medical Academy, Department of Toxicology, Simkova 878, 500 01 Hradec Kralove, Czech Republic
*E-mail: patocka@pmfhk.cz*

The question of aluminum's relevance to the etiology of Alzheimer's disease cannot yet be adequately answered. The mechanisms of how aluminum could evoke the hallmarks of AD are not known. Reflecting many studies, which utilize aluminofluoride complexes in laboratory investigations, we suggest that these complexes may act as the initial signal stimulating impairment of homeostasis, degeneration, and death of the cells. Aluminum ions in the presence of fluoride can accelerate the aging and impair the functions of the nervous system. In respect to the etiology of Alzheimer's disease, the long-term synergistic action of aluminum ions and fluorides may represent a hidden but serious and powerful risk factor for the development of this devastating threat to human civilization.

**Keywords:** Alzheimer's disease, Aluminum, Fluoride, Aluminofluoride complexes, G-protein

## List of Abbreviations

| | |
|---|---|
| 1,4,5-$IP_3$ | inositol 1,4,5-trisphosphate |
| 1-IP | inositol 1-phosphate |
| AD | Alzheimer's disease |
| AChE | acetylcholinesterase |
| $AlF_x$ | aluminofluoride complex |
| AMP | adenosine monophosphate |
| ApoE | apolipoprotein E |
| APP | amyloid-precursor protein |
| ATP | adenosine triphosphate |
| BBB | blood-brain barrier |
| $[Ca^{2+}]_i$ | cytosolic calcium level |

| | |
|---|---|
| cAMP | cyclic AMP |
| CMP | cytidine monophosphate |
| DAG | 1,2-diacylglycerol |
| GAP | guanosine triphosphatase activating protein |
| GDP | guanosine bisphosphate |
| G-protein | guanosine triphosphate-binding protein |
| GTP | guanosine triphosphate |
| GDP(S) | guanosine 5′-[$\beta$ thio]diphosphate |
| GTP(S) | guanosine 5′-[thio]triphosphate |
| IPs | inositol phosphates |
| LPA | lysophosphatidic acid |
| PA | phosphatidic acid |
| PI | phosphatidylinositol |
| PIP | phosphatidylinositol 4-phosphate |
| $PIP_2$ | phosphatidylinositol 4,5-bisphosphate |
| $PLA_2$ | phospholipase $A_2$ |
| PLC | phospholipase C |
| PLD | phospholipase D |
| PKC | protein kinase C |
| RBC | red blood cells |
| TSH | thyroid-stimulating hormone |

## 1 Introduction

Alzheimer's disease (AD) is a neurodegenerative disorder with impairment of cognitive functions and the loss of memory. The number of patients has risen dramatically in the industrialized world as well as in the developing countries. AD is among the most frequent obstacles to healthy aging. The increasing size of the very old population has led to the increasing impacts on the cost of social and health care. In spite of accumulated knowledge of cellular and molecular aspects, it is not known how to arrest or delay specifically the course of this devastating disease. Intense efforts are underway to find a cause or a systemic metabolic disorder, which could be targeted by therapy.

The idea that aluminum is the risk factor for AD has attracted considerable attention from general public as well as from researchers. A higher amount of aluminum was found in the human brain with AD than in brains of age-matched healthy controls [1, 2]. The hypothesis that the accumulation of aluminum in the brain is the cause of this disease has been postulated and discussed very often [3–5]. This suggestion was further supported by a positive correlation between the incidence of AD and concentrations of aluminum in drinking water. Nine out of 13 published epidemiological studies of aluminum drinking water and AD have shown statistically significant positive relations [6]. The study including 3,777 subjects aged 65 years in southwestern France suggests that a concentration of aluminum in drinking water above 0.1 mg per liter may be a risk factor of dementia and, especially AD [7]. A major problem in the interpretation of these studies is that drinking water only contributes a fraction of the total dietary intake of aluminum. Yokel [8] calculated that drinking water provides only about 1% of normal daily human intake. Only some of the studies that determine aluminum in post-mortem brains, senile plaques and neurofibrillary tangles of patients with AD are consistent with the aluminum hypothesis. The relatively poor sensitivity of some of the techniques available for these analyses could explain the discrepancies [9]. Considerable evidence exists that aluminum may play a role in the etiology and pathogenesis of AD, but whether the link is causal is still open to debate.

Recent fundamental research of the pathogenesis of AD brings evidence that this disease is connected with the alterations in neurotransmission, abnormal polymerization of cytoskeletal protein $\tau$, changes in $\beta$-amyloid precursor protein and $\beta$-amyloid production, apolipoprotein E accumulation, and alterations in mechanisms of calcium homeostasis. Inflammation and impairment in the action of free oxygen radicals are probably also involved in the pathogenesis of AD [10]. Three genes have been identified that cause the less common early-onset, familial cases of the disease: the $\beta$-amyloid precursor protein gene on chromosome 21, the presenilin 1 gene on chromosome 14, and the presenilin 2 gene on chromosome 1. More than 50% of the cases are late-onset and related to the apolipoprotein E gene on chromosome 19 [11].

The mechanisms of how aluminum could evoke the multiple pathophysiological changes in AD are not known. Many authors suggest that aluminum alone is not a cause of AD. Neither the increased content of aluminum in the brain nor environmental studies can explain whether and why aluminum constitutes a risk. Aluminum is therefore regarded as a putative risk factor along with alcohol consumption, stressful life events, and manual occupation. On the other hand, a comprehensive book reflecting the myriad of ways that aluminum is known to impact upon mammalian physiology and the etiology of AD has been published recently [10].

## 2 Aluminofluoride Complexes

The intensive laboratory research of mechanisms of signal transduction brings numerous experimental data, which could change our understanding and interpretation of aluminum's action at the cell level. Reflecting many laboratory studies we suggest that some of the pathological changes are not raised by aluminum alone, but by the synergistic action of aluminum and fluoride [12].

Fluoride anions have long been known to influence the activity of various enzymes and adenylate cyclase [13]. A breakthrough came later when Sternweis and Gilman [14] demonstrated that fluoride activation of adenylate cyclase depends on the presence of traces of aluminum. This fact had at first been ignored because aluminum was present as a leached-out impurity in millimolar solutions of fluoride in glass. Gilman and coworkers found that the target of the activation was a heterotrimeric G-protein and that the active stimulatory agent was aluminum fluoride. The soluble aluminofluoride complexes ($AlF_x$) are formed in water solutions containing fluoride and traces of aluminum [15, 16]. The idea that the aluminofluoride complex acts as a new high affinity analogue of $\gamma$-phosphate [16, 17] has been accepted during the last decade and a great number of reports on its use has appeared.

## 2.1 The Structure of Aluminofluoride Complexes

The aluminofluoride complexes are not permanent and the proportions of species such as $AlF_3$ and $AlF_4^{1-}$ are still disputed. Phenomenological observations seems to verify that the pH determines the complexation state of $AlF_x$. Factors other than pH, such as fluoride concentration and the location of positively charged amino acid of the active site of the phosphoryl-transferring enzyme, may cause deviation from the strict pH dependence of $AlF_3$ versus $AlF_4^{1-}$ in biological systems [18]. In aqueous solutions with a pH of less than 5.5, aluminum exists as the octahedral hexahydrate $Al(H_2O)_6^{3+}$, usually abbreviated to $Al^{3+}$. At pH values above 6.2 $Al(H_2O)_6^{3+}$ undergoes successive deprotonation, becoming tetrahedral aluminate $Al(OH)_4^{1-}$ [19]. If fluoride is added, the four equatorial water molecules of the hexahydrate are replaced by fluoride to give $AlF_4(H_2O)_2^{1-}$. At pH values above 6.2 the tetrahedral aluminate species predominates with a varying composition of hydroxy groups and fluoride.

Chabre and coworkers [16, 17] suggested that $AlF_4^{1-}$ is probably the active species, which mimics the role of the $\gamma$-phosphate. According to these authors, it is unlikely that the whole $AlF_4^{1-}$ complex binds with its four fluoride atoms. Most probably, when entering the site, one of the fluoride atoms is released and the tetracoordinated aluminum binds to the oxygen on the $\beta$-phosphate [16, 17].

What the high concentration of fluoride in the solution does is to induce the formation of a soluble tetracoordinated state of aluminum, which has the same geometry, size, and coordination as a phosphate. Among many metals tested in the presence of fluoride, Sternweis and Gilman [14] found that only beryllium could substitute for aluminum for the activation of adenylate cyclase. This fact reinforced the assumption that aluminum acts through its tetrahedral phosphate-like complex $AlF_4^{1-}$, because all beryllium complexes are tetracoordinated [16].

Martin [19] has recalculated the equilibria of $AlF_x$ and contested that $AlF_4^{1-}$ is the major species in the presence of millimolar fluoride at neutral pH and is tetrahedral in aqueous solution. According to his calculations, the predominant species are the neutral complex $AlF_3$ and the mixed complex $AlF_3(OH)^{1-}$. These complexes should be hexacoordinated, with water molecules occupying the free sites. It is uncertain whether the complex that enters the site is an $AlF_3$ that becomes tetrahedral by losing its three bound water molecules and contracting a fourth bond with the $\beta$-phosphate oxygen, or if it is an already tetrahedral $AlF_3(OH)^{1-}$ that exchanges its hydroxyl for the $\beta$-phosphate oxygen or an $AlF_4^{1-}$ that exchanges a fluorine. In earlier reviews Martin had also admitted that the active species might be simply the aluminum ion $Al^{3+}$ and that fluoride complexation would only be needed to allow the penetration of aluminum across the plasma membrane. Once corrected for these effects, Antonny and Chabre [20] proposed that $AlF_3(OH)^{1-}$ is the main activating species and that the bound form of the complex is tetracoordinated GDP-$AlF_3$. These complexes can bind to proteins by hydrogen bonds to the fluorine atoms just as to oxygen atoms of a phosphate ion. So can arise an aluminofluoride analogue of pyrophosphate, R-O-$PO_2$-O-$AlF_3$, which may be bound at the site for the $\gamma$-phosphate. Thus, for instance, in G-protein whose nucleotide site already contains GDP, $AlF_x$ can form hydrogen bonds with the donor groups of the $\beta$-phosphate site and bind ionically to the terminal oxygen of the $\beta$-phosphate of the GDP. $AlF_x$ was found to be a good analogue of a $\gamma$-phosphate for a number of ATP- and GTP-converting enzymes. Enzyme-bound GDP or ADP could therefore form a complex with $AlF_x$ that imitates ATP or GTP in its effect on protein conformation.

Studies of the crystal structures of nucleotide binding proteins complexed with $AlF_x$ show that in the case of G-proteins, myosin-S1, and nitrogenase, $AlF_4{}^{1-}$ is the active site species, whereas in others metabolic enzymes, e.g., nucleoside diphosphate kinase and uridylate monophosphate kinase [18, 21], $AlF_3$ is bound. It is not clear whether these differences reflect some differences between proteins or are due to technicalities. Schlichting and Reinstein [18] compared the coordination numbers at different crystallization conditions and suggested that the different coordination numbers originate mainly from the difference in pH at which the enzymes were crystallized. $AlF_x$ occurs as $AlF_3$ at pH 7.5–8.5 but as $AlF_4{}^{1-}$ at pH below 7. To minimize the influence of factors other than pH on manifestation of $AlF_x$, they treated the two uridylate monophosphate kinase-ADP-CMP-$AlF_x$ complexes (this enzyme from *Dictyostelium discoideum* catalyzes the reversible trans-phosphorylation between ATP and CMP) as similarly as possible except for the pH of the crystallization buffer. The bound $AlF_x$ occurred as $AlF_3$ at pH 8.5 but as $AlF_4{}^{1-}$ at pH 4.5.

## 2.2
## Aluminofluoride Complexes and G-Proteins

Chabre and coworkers [16, 17] suggested that $AlF_4{}^{1-}$ mimics the role of the $\gamma$-phosphate only if the $\beta$-phosphate is present and remains unsubstituted on its oxygen. The effect is more readily seen with G-proteins because GDP is always tightly bound in the nucleotide site of the protein. Low cost and availability of $AlCl_3$ together with NaF has probably contributed to their widespread use as a tool in laboratory studies of G-proteins [12]. These studies brought a great deal of knowledge about the involvement of heterotrimeric G-proteins in cell signaling. The heterotrimeric G-proteins mediate the transfer of information from cell-surface receptors to effector molecules. Heterotrimeric G-proteins thus play a pivotal role in transmembrane signaling processes.

### 2.2.1
### *G-Protein-Coupled Receptors*

The family of cell-surface receptors that require coupling to G-protein transducers for functional signaling is vast and diverse. These receptors detect inputs, such as photons, odorants, neurotransmitters and hormones, such as dopamine, epinephrine, norepinephrine, serotonin, acetylcholine, glucagon, vasopressin, melatonin, neuropeptides, opioids, excitatory amino acids, prostanoids, purines, growth factors, and morphogens, linking extracellular stimuli to intracellular signaling networks. The isolation of cDNAs for G-protein-coupled receptors enabled a comparison that indicated a shared structural configuration. These receptors consist of a single protein chain that crosses the plasma membrane seven times. The cytosolic C-terminal domain, Asp-Arg-Tyr (339–341) motif at the cytosolic end of helix III, and the loops between the helices III–IV and V–VI are involved in G-protein recognition and binding [22].

### 2.2.2
### *G-Proteins*

The members of heterotrimeric G-protein family are composed of a $\alpha$-(39–52 kDa), a $\beta$-(35–36 kDa), and a $\gamma$-(8–10 kDa) subunit. The G-protein $\alpha$-subunit binds to GDP in the inactive state. The agonist receptor binding facilitates the exchange of GDP for GTP. The activated $\alpha$-subunit dissociates from $\beta$- and $\gamma$-subunits to interact with effector enzymes on the cytosolic side. Low $K_M$ GTPase intrinsic activity of the $\alpha$-subunit hydrolyzes GTP to GDP to end the cycle activation [23]. Simultaneously, a conformational change occurs, leading to an increased GTPase activity of the $\alpha$-subunit. On washing off the agonist, the GTP-bound form converts to the GDP-bound form, which is dependent on the rate of GTPase activity. GTP hydrolysis drives the conformational change and is responsible for interrupting the interaction with the effector molecule. As G-proteins are activated when they go from the GDP-bound to the GTP-bound state, it was assumed that the $AlF_x$ occupies the $\gamma$-phosphate-binding site on the protein and together with bound GDP makes the G-protein act as if it has bound GTP. Chabre [17] concluded that $AlF_x$ does not seem to form an analogue of the transition state since the structural change locks the site and prevents the dissociation of the analogue. The determination of the three-dimensional structures of heterotrimeric G-proteins bound to GDP and $AlF_x$ in 1994 unexpectedly demonstrated that they mimic the transition state of the phosphoryl-transfer reaction rather than the ground state. This is because aluminum is bound to four fluoride ligands in a square-planar coordination with two oxygen ligands at the apical positions of the resulting octahedron [21]. One oxygen ligand is a $\beta$-phosphate oxygen, whereas the other is the oxygen from water believed to represent the attacking nucleophile of the hydrolysis reaction. Invariant arginine and glutamine residues, which are required for GTP hydrolysis but not for its binding, stabilize the $AlF_x$ binding to GDP. In addition, a new class of GTPase-activating proteins (GAPs) has been discovered to modulate the rate of G-protein-mediated effects. Biochemical evidence showed that GAPs (also called regulators of G-protein signaling) bind with higher affinity to G · GDP · $AlF_x$ complex of G α-subunit than to the triphosphate state of G-protein. GAPs also stabilize the GTPase active conformation of G-proteins. These observations further support that $AlF_x$ stabilizes the transition state [24].

The observation that $AlF_x$ complexes activate G-proteins has been therefore useful for the study of the mechanism of G-protein activation, for understanding the biochemical mechanism of GTP hydrolysis, and for the elucidation of three-dimensional structures of several GTPases, including the discovery of the GTPase-activating proteins.

Although it was initially believed that $AlF_4{}^{1-}$ only interacts with Gα-subunits, recent studies have revealed that a small G-protein Ras can interact stably with $AlF_3$ in the presence of RasGAP [24]. The proto-oncogene product Ras is a component of intracellular signaling pathways involved in cell growth and division. It has a very low intrinsic GTPase reaction rate that is stimulated 105-fold by RasGAPs that down-regulate the accumulation of Ras · GTP. The determination of the structure of a complex between RasGAP and Ras · GDP in the presence of aluminum ions and fluoride [21, 24] shows that $AlF_3$ forms a pentagonal bipyramid, with the fluorides forming the trigonal base with two apical oxygen ligands. Similar studies demonstrated that several classes of small GTPases can stable interact with their respective GAPs in the presence of $AlF_3$, suggesting that the aluminofluoride complex could bind to a wide variety of GTPases. These observations demonstrate that the GTP hydrolysis mechanism is similar for both small GTPases and Gα-subunits.

In most systems, the α-subunit determines the specificity of interaction with receptor and effector molecules. Molecular cloning has revealed a large molecular diversity within the subunits of G-protein. During the last decade, the number of known mammalian G-protein α-subunits grew rapidly to comprise more than 20 individual proteins. There are 5 G-protein β-subunits and at least 12 γ-subunits [25]. Small monomeric G-proteins were found in eukaryotes from yeast to human. They constitute a superfamily consisting of more than 100 members, classified into at least five families: the Ras, Rho, Rab, Sar1/Arf, and Ran. The Ras family regulates gene expression, the Rho family regulates cytoskeletal reorganization and gene expression, the Rab and Sar1/Arf families regulate vesicle trafficking, and the Ran family regulates nucleocytoplasmic transport and microtubule organization. Many upstream regulators and downstream effectors of small G-proteins have been isolated, and their modes of activation and action have gradually been elucidated [26].

### 2.2.3
### *G-Protein-Coupled Effector Enzymes*

Adenylyl cyclase is the effector enzyme that synthesizes cyclic adenosine monophosphate (cAMP) from ATP. Cyclic AMP functions as a second messenger to relay extracellular signals to intracellular enzymes such as protein kinase A. Adenylyl cyclases are integral membrane proteins that consist of two bundles of six transmembrane segments. Two catalytic domains extend as loops into the cytoplasm. There are at least nine isoforms of adenylyl cyclase, based on cloning of full-length cDNAs. Early studies indicated that cyclase activity was regulated primarily by interactions with $\alpha$-subunits of heterotrimeric G-proteins. Binding of a stimulatory $\alpha$-subunit ($G_s$) enhanced activity while binding of an inhibitory $\alpha$-subunit ($G_i$) inhibited adenylyl cyclase activity [23, 26]. More recently, it has become clear that cyclase activity is regulated by multiple effectors, which include not only the $\alpha$-subunits of $G_s$ and $G_i$ proteins, but also the G-proteins $\beta\gamma$-subunits and the protein kinase C. Five of the adenylyl cyclases known are regulated by calcium [27].

Phospholipase C (PLC) catalyzes the phosphodiesteric cleavage of phosphatidylinositol 4,5-bisphosphate ($PIP_2$) into two moieties, inositol 1,4,5-trisphosphate (1,4,5-$IP_3$) and 1,2-diacylglycerol (DAG) [27, 28]. 1,4,5-$IP_3$ is a second messenger that controls many cellular processes by generating intracellular calcium signals. The other second messenger DAG is responsible for activating a protein kinase C (PKC), which catalyzes protein phosphorylation. $PIP_2$ is the main substrate for PLC. Many isoforms of PLC have been identified, based on their purification and characterization, and upon the established amino acid sequence. The differing functions of the various PLC families are not yet fully known. Some isoforms are regulated by G-proteins and can, in turn, regulate these by serving as GTPase-activating G-proteins (e.g., PLC$\beta$1). The $G\alpha_{q/11}$-subunit activates $\beta$-isoforms of PLC. Others are regulated by tyrosine-kinases (e.g., PLC $\gamma$) [28].

Tyrosine kinase receptors generate 1,4,5-$IP_3$ and DAG by interacting directly with PLC$\gamma$. PLC$\gamma$ contains two $SH_2$ domains and one $SH_3$ domain. The $SH_2$ domain of PLC$\gamma$1 recognizes and binds to a specific phosphotyrosine residue. Once PLC$\gamma$1 is phosphorylated, it starts to hydrolyze $PIP_2$. Signal transduction using tyrosine kinase receptors is an energy-consuming process: ATP is utilized for receptor phosphorylation and for PLC$\gamma$ phosphorylation. The resynthesis of phosphoinositides also requires ATP for the phosphorylation of phosphatidylinositol [27].

Phospholipase D (PLD) is now known to be also involved in receptor-effector cascade [29]. Activation of PLD occurs concurrently with PLC activation. This enzyme is stimulated by receptors having tyrosine kinase activity. PLD preferentially hydrolyzes phosphatidylcholine to yield phosphatidic acid (PA) and choline. PA is hydrolyzed by PA phosphohydrolase to produce DAG. DAG might be in turn phosphorylated by DAG kinase to yield PA. PA may also serve as a potential lipid messenger, or it undergoes attack by phospholipase $A_2$ ($PLA_2$) to yield lysophosphatidic acid (LPA). LPA has been described as the lipid mediator, which acts through a specific G-protein-coupled receptor [30]. When LPA is added, PLC is activated and adenylate cyclase is inhibited through $G_i$. LPA stimulates PLD activity and also induces arachidonate release, probably as a result of $PLA_2$ activation. $PLA_2$ may generate arachidonic acid, which goes on to form prostaglandins, thromboxanes and lysophospholipids. $PLA_2$ was found in the cytosol of non-stimulated cells. $Ca^{2+}$ binding in the lipid domain and phosphorylation enhances the translocation of $PLA_2$ to the plasma membrane surface. It has been suggested that G-proteins also mediate the activation of $PLA_2$.

The evidence is compelling that ion channels are direct G-protein effectors. $K^+$ channels are activated by $G_a$ and inhibited by $G_i$, whereas the opening of L-type $Ca^{2+}$ channels is regulated by $G_s$ or $G_o$ protein [31]. Studies on G-protein-ion channel membrane interactions have shown that G-proteins may have several membrane targets, including more than one type of ionic channels.

## 2.3 Aluminofluoride Complexes in Laboratory Investigations

Fluoride activation in the presence of trace amounts of aluminum has been often used as evidence for involvement of a G-protein in a system. A great number of reports on their use has appeared, with far-reaching consequences for our understanding of fundamental living processes. Laboratory investigations during the past decade bring evidence that aluminofluoride complexes influence all cells and tissues of the human body with powerful pharmacological efficacy [12]. It has been demonstrated that aluminum ions in the presence of millimolar fluoride may clone or potentiate the action of numerous extracellular signals. The principle of amplification of the initial signal during its conversion into functional response has become a widely accepted tenet in cell physiology (Fig. 1).

$AlF_x$ therefore affects the levels of second messenger molecules and cytosolic calcium with extensive effects on cell metabolism [12, 32]. Biological signaling pathways interact with one another to form complex networks. The question as to how cells manage to make the right responses at the right times has yet to be fully understood. The discoveries of receptor diversity, numerous isoforms of G-proteins and effector molecules broaden enormously the possibilities of interactions of signal transduction events. Yet, it seems that we shall not probably find any physiological process, which is not potentially influenced by $AlF_x$. Science has already accumulated evidence how diverse biological problems can be investigated using the small inorganic molecule $AlF_x$ [12, 21, 32]. It has led to the advance in our understanding of living processes, the common denominator of which is the transfer of a phosphoryl group. The phosphate-analogue model of $AlF_x$ action was accepted for G-proteins but it was later extended to all enzymes that bind phosphate or nucleoside-polyphosphate. Phosphoryl-transfer reactions are involved in processes such as cell signaling, energy transduction, regulation of cell metabolism, cytoskeletal protein assembly, growth, cell differentiation, aging, and apoptosis. Considering that these reactions are fundamental for nearly all biological systems, we can understand the choice of aluminum fluoride as the biological molecule of the month March 1997 [21]. In view of the ubiquity of phosphate in cell metabolism, a phosphoryl-transfer transition state analogue might represent a useful tool for laboratory investigations, but also a strong potential danger for living organisms including humans. We should try to understand how the simple signal could lead to multiple events resulting into a pathophysiological state.

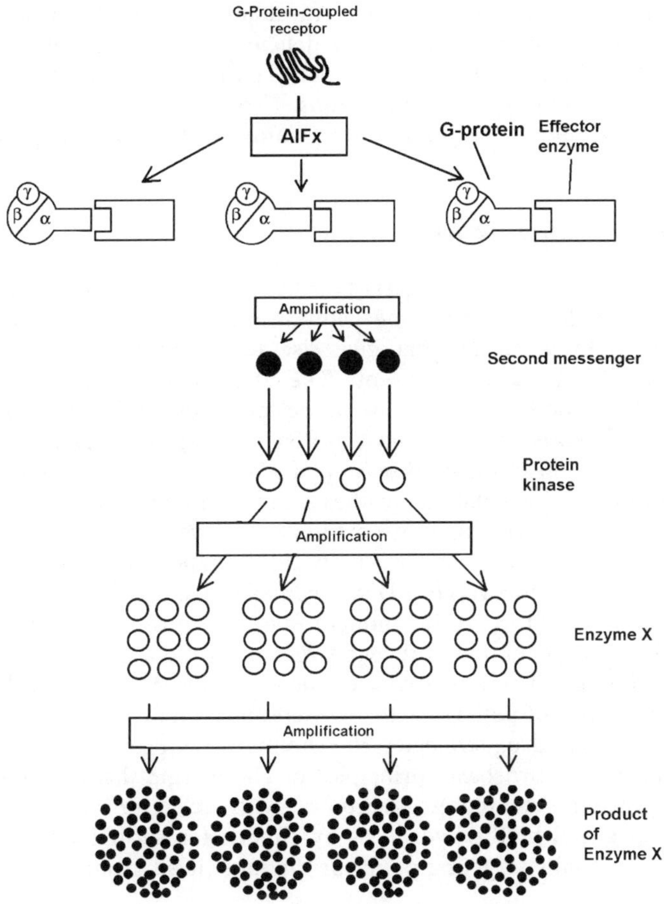


**Fig. 1.** $AlF_x$ acts as the messenger of a false information. Its message is greatly amplified during the conversion into the functional response of a cell. The second messenger molecule could be cAMP, 1,4,5-$IP_3$, and DAG. Moreover, $AlF_x$ can participate as the analogue in the phosphoryl-transfer reactions involved in the signaling cascade

## 3 Reevaluation of the Role of Aluminofluoride Complexes in the Etiology of Alzheimer's Disease (AD)

The scientific world and the pharmaceutical companies are seeking one reaction, one enzyme, one substrate or one gene, which could be targeted by therapy. The biochemical approach is based on the hypothesis that the causative disorder of AD may lurk at the level of cell physiology, altering various brain processes, such as the recognition, memory, language, and visuospatial skills. The recent fundamental research of pathogenesis of AD has brought evidence that this disease is connected with the alterations in neurotransmission, $\beta$-amyloid production, plaque formation, and cytoskeletal abnormalities in brain tissue [5, 10]. An alteration in calcium homeostasis could affect most of these AD-related deficiencies. A disturbance in signal transduction between brain cells could underlie the above-mentioned AD-related pathophysiological alterations. G-proteins and transfer of phosphate groups are the most important regulatory factors in these processes.

## 3.1
## Cholinergic Transmission and $AlF_x$

The cholinergic hypothesis suggests that the reduction of cholinergic neurotransmission could explain the most important cognitive deficit in AD [33]. The basal forebrain nuclei that give rise to cholinergic fibers show cell degeneration in brains of AD patients. The 60–70% depletion of acetylcholinesterase and choline acetyltransferase in the cortex and in the limbic structures of patients with AD is one of the most consistent and profound changes as early as one year after onset of the disease [34, 35]. The cortex of patients shows decreased rates of acetylcholine synthesis and choline uptake. Aluminum has been found to inhibit choline transport in isolated rat brain nerve endings and reduce neuronal acetyltransferase activity [5]. It is therefore hypothesized that raised brain aluminum levels in AD may contribute to the cholinergic neuronal deficits in AD. Therefore, it was thought that the study of cholinergic lesions could explain the most important cognitive deficits and help in the discovery of a rational therapy. The alterations of nicotinic and muscarinic cholinergic receptors in brains of patients with AD were found [36].

Among different drug strategies, the cholinergic approach has gained great interest. There are two basic principles of cholinergic therapy: first, stimulation of various cholinergic receptors with selective agonists; second, reduction of acetylcholine hydrolysis by cholinesterase inhibitors. Cholinesterases are today the targeting point for the most used group of AD drugs.

### 3.1.1
### *Cholinergic Receptors*

Molecular biology shows the diversity of acetylcholine receptors. Twelve genes were found for nicotinic cholinergic receptors; the synthesis of muscarinic receptors is coded by five separate genes [37]. Different subtypes of these receptors have different physiological and pharmacological characteristics. Muscarinic cholinergic receptors m1, m3, and m5 stimulate the generation of cAMP, but in the case of m1 such stimulation is secondary via the activation of phospholipase C, production of inositol 1,4,5 trisphosphate and calcium mobilization. These receptors also stimulate the activation of $PLA_2$, PLD, and tyrosine kinase, and activate voltage-gated calcium channels. On the other hand, muscarinic receptors m2 and m4 inhibit the activity of adenylate cyclase, calcium influx, and the activity of $PLA_2$. The cholinergic receptors in the brain are mostly muscarinic and their activation has been connected with the cognitive function. At postsynaptic muscarinic receptor sites acetylcholine acts by reducing potassium conductance making the cholinoreceptive neuron more susceptible to other excitatory inputs [38]. The discovery of different muscarinic receptor subtypes has offered new opportunities to develop specific inhibitors of the m1 postsynaptic receptor subtypes. However, the therapeutic strategy of direct pharmacological activation of muscarinic receptors is somehow questionable since the additional effects could be derived from muscarinic receptor antagonism. It is unclear, e.g., whether treatment should aim to enhance or diminish glutamatergic transmission, since the positive modulation could facilitate learning while increasing the glutamate function may enhance excitoxicity and neuronal death.

The decreased number of nicotinic acetylcholine receptors in the cortex and hippocampus associated with AD and the beneficial effects of nicotine observed in AD patients suggest the important role played by nicotinic receptors [39]. Presynaptic nicotinic receptors have a key role in the regulation of release of many transmitters such as acetylcholine, glutamate, and serotonin. Indeed, AD involves more than cholinergic deficiency. In brains of patients with AD several other transmitters are depleted. Attempts to intervene in acetylcholine receptors pharmacologically may thus present us with large problems. For example, agonists of muscarinic receptors stimulate the formation of $\beta$-amyloid whose presence in the membranes induces the uncoupling of muscarinic receptors from phospholipase C. Stimulation of muscarinic acetylcholine receptors leads to phosphorylation of $\beta$-amyloid precursor protein, promoting the plaque formation.

Similar interactions may occur in the presence of the small molecule $AlF_x$. The acetylcholine receptors were not thought of as G-protein receptors, but the use of fluoride plus aluminum salts contributed to the discovery of very complicated regulatory and modulatory relationships in their action [31].

The use of $AlF_4^{\,1-}$ in the investigation of canine cerebral cortex [40] led to the discovery of a previously unrecognized signaling pathway in the brain. These experiments demonstrated that muscarinic acetylcholine receptor-G-protein uses PLD as the effector enzyme. $AlF_4^{\,1-}$ caused a two- to three-fold increase in breakdown of phosphatidylcholine and rapid accumulation of choline and phosphatidic acid was observed.

Binding studies of muscarinic-receptor agonists suggested that mechanisms of G-protein-receptor coupling or uncoupling might be disturbed in AD [41]. Wallace and Claro [42] reported the increased sensitivity of PLC to GTP(S) during AD. Also the impairment of carbachol-stimulated low $K_m$ GTPase activity in AD basal ganglia was observed. However, this area warrants further study. In view of the role of both heterotrimeric as well as small G-proteins, the aluminofluoride complex may act as a bullet, which could evoke the marked disturbances in complex networks in the brain.

### 3.1.2
### *Acetylcholinesterase*

Detailed studies of the molecular structure of acetylcholinesterase (AChE) have shown that this enzyme exists in many molecular forms [43]. There are globular (G) and asymmetric classes, which are further subdivided. The globular forms predominate in mammalian brain. Most brain AChE consist of assemblies of four identical 77 kDa catalytic subunits linked by disulfide bridges ($G_4$). The brain contains a small amount of the monomeric $G_1$ form with a small amount of $G_2$ form. The only AChE molecular form found in erythrocytes is globular dimer $G_2$. The enzyme is anchored in the membrane by a glycosylated phosphatidylinositol. Despite this multiplicity, all cholinesterase molecular forms have equivalent catalytic activities, but some inhibitors affect them differentially. The decrease of presynaptically bound $G_4$ and the increase of the soluble $G_1$ a $G_2$ form was observed in the brains of aging rats [38]. Similar changes were found in the frontal and parietal cortex, in the hippocampus and cholinergic projection nuclei in brains of AD patients.

Platt and coworkers [44] applied histochemical techniques to identify changes in AChE distribution induced by intracerebroventricular aluminum injections (5.4 μg in 5.5 μL, daily over a period of 5 successive days) in the adult rat brain after survival periods of either 1 or 6 weeks. Damage of the cingulate bundle in aluminum-treated animals led to a severe anterograde degeneration of cholinergic terminals in the cortex and hippocampus, as indicated by AChE labeling. Immunoreactivity of astrocytes and phagocytic microglia revealed a greater inflammatory response in aluminum-injected animals compared to controls. These data suggest that the enhancement of inflammation and the interference with cholinergic projections may contribute to pathological processes in AD.

Therapeutic application of AChE inhibitors requires the need to monitor the activity of this enzyme on the periphery. Since the brain is the target organ for all potential cholinergic drugs, any peripheral measures can provide important information about a compound efficacy and mechanism of action. Such studies are relatively non-invasive and simple. The reliable correlation between peripheral and central cholinesterase inhibition in humans depend on many factors, clearly vary from drug to drug, and require detailed pharmacokinetic studies.

#### 3.1.2.1
#### The Effect of $AlF_x$ on AChE Activity in Human Red Blood Cells

We studied the effect of 20 μM $AlCl_3$ and 0.01–1 mmol $L^{-1}$ NaF on the AChE activity in erythrocytes of AD patients, age-matched healthy subjects, and young healthy controls. The AD patients were chosen on the level of clinical diagnosis "probably Alzheimer's disease" according to DSM IV, NINCDS/ADRDA criteria, Hachinski Ischemic Score, GBS scale, and MMSE. No subject of this study had concurrent somatic illness or a neurological or psychiatric disorder. The study was approved by the Ethic Committee of the Prague Psychiatric Center, and before admission to the study a written informed consent was obtained from all subjects. While 20 μM $AlCl_3$ increases the AChE activity in RBC in all groups of investigated subjects to 158–285%, 1 mM NaF and 20 μM $AlCl_3$ inhibit the AChE activity in intact RBC as well as in hemolyzate in all groups of investigated subjects (Table 1).

**Table 1.** The effect of 1 mM NaF + 20 μM $AlCl_3$ on the acetylcholinesterase activity (AChE) in freshly prepared intact RBC and in hemolysate of patients with AD (mean age 72.5 ± 5.1 years), age-matched healthy controls (AM-HS) (72.1 ± 1.6 years), and the group of young healthy subjects (YS) (35.9 ± 8.5 years). Whole venous blood samples were drawn from each subject after overnight fasting., always at 07:30 AM. Red blood cells (RBC) were isolated from the blood of patients with AD, AM-HS, and YS by centrifugation [68]. RBC AChE activity was evaluated in intact freshly prepared RBC or hemolyzate following the spectrophotometric method [45] with modifications. Buffer was Tris-HCl, pH 7.5 in the solution of 154 mmol $L^{-1}$ NaCl, acetylthiocholine iodide was a substrate. Measurement of enzymatic activity was performed in fluorimeter polystyrene cuvettes for 3 min (UV/VIS spectrophotometer Shimadzu, Japan). The effects of 1 mmol $L^{-1}$ NaF in the presence of 20 μmol $L^{-1}$ $AlCl_3$ were measured. Data are expressed in percentage of the AChE activity in the absence of aluminum and fluoride ions. No differences between the AChE activity were found between the investigated groups

| Subjects | AD | AM-HS | YS |
|---|---|---|---|
| N | 23 | 16 | 16 |
| Intact RBC | 16.3 ± 5.5 | 18.9 ± 4.2 | 17.1 ± 2.3 |
| Hemolyzate | 19.8 ± 3.5 | 21.4 ± 4.6 | 26.0 ± 6.3 |

While 20 μmol $L^{-1}$ $AlCl_3$ increased the AChE activity in RBC in all groups of investigated subjects to 158–285%, in the presence of 1 mmol $L^{-1}$ NaF aluminum inhibited the AChE activity in intact RBC as well as in hemolyzate in all groups of investigated subjects (Table 1). The inhibition induced by $AlF_x$ was dependent on the concentration of fluoride in the cuvette (Fig. 2). The activation of AChE activity by 3.7 μmol $L^{-1}$ $Al^{3+}$ in the human RBC membranes and by 1–100 μmol $L^{-1}$ $Al^{3+}$ in bovine brain was observed by Zatta et al. [46].

We could hardly accept these results as the discovery of a new available AChE inhibitor. Therapeutic strategies to affect one enzyme pharmacologically may be followed by various side effects, since AChE is present in many tissues. For example, the initial studies of the effect of tacrine showed its hepatotoxicity. The AD patients with two ε4 alleles of the ApoE gene do not respond to such a therapy and have higher AChE activity in cerebrospinal fluid than controls and AD patients with one or no ε4 [47]. The therapeutic use of AChE inhibitors directed the research on AD to the study the catalytic function of this enzyme. However, the presence of AChE in the RBC plasma membrane does not seem to be connected with the processes of neurotransmission. In connection with the pathogenesis of AD the non-catalytic and non-specific role of AChE is under discussion. AChE co-localizes with $\beta$-amyloid deposits of AD brains [48]. AChE accelerates $\beta$-amyloid formation and it may therefore act as a pathological chaperone inducing a conformational transition of the $\beta$-amyloid protein. This action of AChE was not affected by edrophonium, an active site inhibitor, but it was affected by propidium, a peripheral anionic binding site ligand. Moreover, AChE associated with plaques, tangles and $\beta$-amyloid angiopathies possesses different enzymatic properties, and, quite possibly, is of a different source as compared with the enzyme associated with

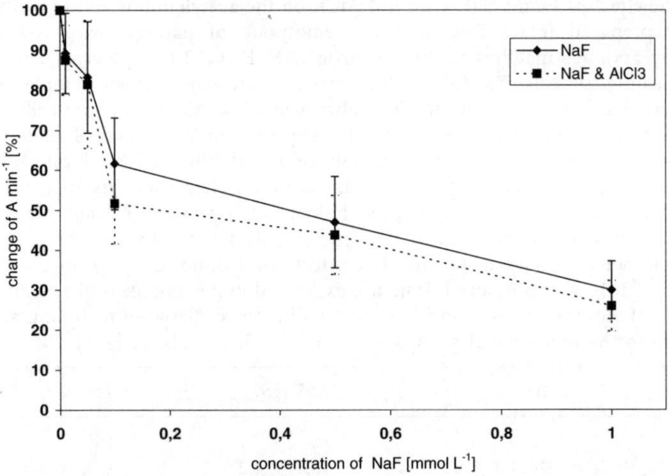


**Fig. 2.** The effect of various concentrations of NaF (-♦-) or NaF in the presence of 20 μmol L$^{-1}$ $AlCl_3$ (-■-) on the AChE activity in hemolyzate from RBC of young healthy subjects. Data are expressed in percentage of the AChE activity in the absence of aluminum and fluoride ions. n = 8–16 measurements

normal neurons and axons [48]. It can thus participate directly not only in the processing of *β*-amyloid protein but also by causing aberrant growth and development of neurons. The results of intensive research of AChE as a therapeutic target provide us with many parts of a puzzle and lead us to postulate many further questions. Is the degeneration of cholinergic neurons and the reduction of acetylcholine the primary change or is it evoked by the action of soluble AChE? Why are there so many forms of AChE? Can erythrocytes release AChE into the brain? Some hypotheses suggest that the pathogenesis of AD may be connected with a systemic cholinesterase abnormality or with damage to the blood-brain barrier, allowing AChE to leak from the brain to the plasma.

## 3.2
## Phosphoinositide Signaling System

During the last decade, a new phosphoinositide signaling system has been postulated [27, 28]. A great amount of experimental evidence supports the general concept that phosphatidylinositol 4,5-bisphosphate ($PIP_2$) from the plasma membrane is hydrolyzed and yields inositol 1,4,5-trisphosphate (1,4,5-$IP_3$) and diacylglycerol (DAG) upon receptor stimulation (Scheme 1). Both products of this hydrolysis catalyzed by phospholipase C (PLC) have second messenger roles. 1,4,5-$IP_3$ binds to a receptor in membranes of the endoplasmic reticulum, which results in a release of $Ca^{2+}$ into the cytosol. The 1,4,5-$IP_3$ receptor comprises an 1,4,5-$IP_3$-gated $Ca^{2+}$-channel. DAG activates protein kinase C (PKC). The coupling between the receptor and PLC is

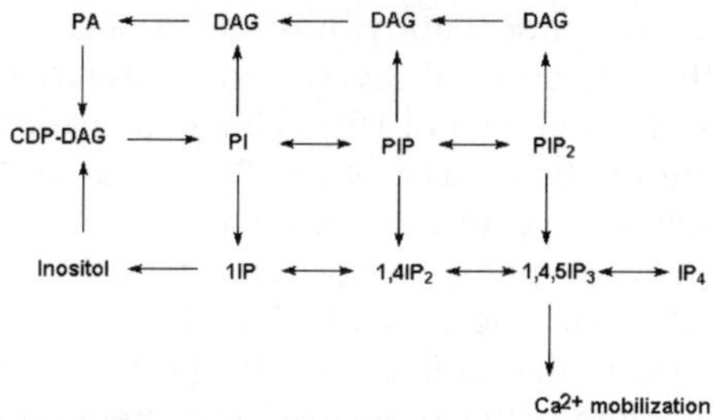


**Scheme 1.** The hydrolysis and resynthesis of phosphoinositides. Hydrolysis of phosphatidylinositol (PI), phosphatidylinositol 4-phosphate (PIP), or phosphatidylinositol 4,5-bisphosphate ($PIP_2$) is catalyzed by PLC in the plasma membrane. The resynthesis of phosphoinositides proceeds in the endoplasmic reticulum

mediated by G-proteins. In this "dual" second messenger hypothesis, 1,4,5-$IP_3$ is the link between $PIP_2$ and $Ca^{2+}$.

Numerous laboratory results demonstrate that micromolar $AlCl_3$ in the presence of millimolar fluoride affects the phosphoinositide turnover, levels of inositol phosphates, and cytosolic free calcium ions in various cells and tissues [32]. Aluminofluoride complexes affect G-proteins involved in the phosphoinositide signaling system [49]. The resynthesis of phosphoinositides and the generation of inositol phosphates requires ATP for phosphorylation. $AlF_x$ might thus also affect the phosphoionositide metabolism as an analogue for a $\gamma$-phosphate for ATP-converting enzymes. Fluoride salts in the presents of aluminum ions have been used to stimulate phosphoinositide hydrolysis in laboratory experiments *in vitro*. The ability of $AlF_x$ to mimic the effects of $Ca^{2+}$-mobilizing hormones suggests that the coupling of hormone receptors to phosphoinositide breaks down through G-proteins [49]. Phosphoinositides might have multiple roles in the cell membranes. For example, phosphatidylinositol serves not only as the parent phospholipid for $PIP_2$ synthesis and the substrate for PLC, but also as the source of arachidonic acid for prostaglandin synthesis [49]. Glycosylphosphatidylinositol serves as lipid anchor of many proteins such as AChE and cytoskeletal proteins [43, 49]. Information on the possible biological significance of this anchoring principle is just starting to emerge, but the implication of inositol lipids in the organization of cytoskeletal proteins can be suggested. Changes in cell-membrane phosphoinositide content appear to result in impaired neurotransmitter-triggered signal transduction related to dysfunction in the coupling of G-proteins to their receptors or effectors.

### 3.2.1
### *The Effect of $AlF_x$ on the Phosphoinositide Signaling System in the Brain*

Phosphoinositides were shown to turn over rapidly in brain and participate in many processes of neurotransmission. The phosphoinositide signaling system is involved in the activation of muscarinic acetylcholine receptors. Early studies on the role of phosphatidylinositol in signal transduction [50] demonstrated that the activation of muscarinic cholinergic receptors resulted in a net loss of phosphatidylinositol (PI). This observation was explained by the initial phosphodiesteratic breakdown of PI, liberating DAG, which was in turn rapidly rephosphorylated in the presence of ATP to yield phosphatidic acid. The involvement of phosphoinositides was later demonstrated for dopamine, $\gamma$-aminobutyric acid and 5-hydroxytryptamine (serotonin) receptors in slices from the guinea pig brain [51]. The discovery of a phosphoinositide signaling pathway focused the attention on the role of $PIP_2$ in agonist-dependent responses [52].

The hypothesis that the stimulated turnover of $PIP_2$ can reflect the increased receptor activation in pathogenic neurons has been postulated [53]. It has been suspected that a change in the amount or in the turnover of inositol lipids can represent a change in the receptor activation since PI or $PIP_2$ is the target of PLC after receptor stimulation.

Chandler and Crews [54] described the formation of inositol phosphates in rat cerebral cortical synaptoneurosomes stimulated by $AlF_4^{1-}$ in sodium-free medium. It was not inhibited by the PKC activator phorbolester, which is known to exert a negative feedback control on agonist-stimulated phosphoinositide metabolism. In experiments with brain cortex membranes, Candura et al. [55] observed that NaF mimicked the action of GTP(S) in stimulating phosphoinositide turnover and generation of inositol phosphates. A much greater hydrolysis of phosphoinositides was observed when $AlCl_3$ and NaF were present together, supporting the concept that $AlF_4^{1-}$ is the active stimulatory species. Coincubation of submaximal concentration of GTP(S) with $AlF_4^{1-}$ did not result in an additive stimulation of phosphoinositide hydrolysis. Paradoxically, $AlCl_3$-induced phosphoinositide hydrolysis was potentiated by coincubation with both GDP(S) and phorbolester. While in hepatocytes $AlF_4^{1-}$ was found to stimulate phosphoinositide metabolism in a manner similar to GTP(S) and sensitive to GDP(S) [56], it seems that $AlF_4^{1-}$ stimulation in brain cortex membranes occurs through a different mechanism [55]. Accumulation of inositol phosphates in the suprachiasmatic nuclei region of rat hypothalamus over a 40 min incubation with $AlF_4^{1-}$ was observed [57].

The tremendous possibilities of potential molecular interactions of aluminum, fluoride and aluminofluoride complexes probably exist in the brain. However, it is evident that interventions of aluminofluoride complexes into the nerve regulations could have severe physiological and pathophysiological consequences.

### 3.2.2
### *Changes in Phosphoinositide Signaling System in the Brains of AD Patients*

The changes in the proportions of various isozymes of PLC with their different specificity to phosphoinositides indicate that the rate of hydrolysis of inositol lipids may be altered in patients with AD. In the normal human brain matter, PLC-$\beta$ was the main type present. In AD brains, the amount of PLC-$\beta$ was significantly reduced in the membranous fraction of the temporal cortical tissues. PLC-$\delta$1 was abnormally accumulated in neurofibrillary tangles, the neurites surrounding senile plaque cores, and neuropil threads in AD brains [58]. The finding that antigenic determinants for PLC-$\delta$1 and PLC-$\gamma$ are shared by several filamentous neuronal inclusions occurring in diverse neurologic disorders may reflect a close association between the phosphoinositide signaling system and cytoskeletal components. In this connection, the aberrant PLC-$\delta$1 accumulation is a common metabolic defect underlying the formation of these inclusions. The total PLC activity was significantly reduced also in platelets of AD patients. These results indicate that a PLC abnormality is present in non-neural tissues as well as in the brains of AD patients. On the other hand, no significant differences in PI-specific PLC activity were found in post-mortem brains between control and AD patients [59]. However, data obtained from the analysis of post-mortem brains of AD patients may have been affected by the long and variable post-mortem delay, because phosphoinositides are rapidly degraded post-mortem. With respect to changes in phosphoinositide metabolism in AD, there is growing evidence that specific changes in this signaling system could be implicated in the development of all characteristic AD lesions.

### 3.3
### The Effect of Aluminum and Fluoride on Calcium Homeostasis

Calcium has been recognized for a long time as an important regulator of many physiological processes. Extensive experimental evidence, accumulated during the last decade, has demonstrated the highly diversified mechanisms in the control of calcium homeostasis. Numerous recent papers have documented that the cytosolic $Ca^{2+}$ level oscillates when cells are exposed to agonists that stimulate hydrolysis of inositol lipids [28]. It also seems, that each cell has its own specific "calcium fingerprint". The calcium transients may serve as an important mechanism encoding information about stimulus intensity and activating selectively different Ca-dependent pathways. It has been demonstrated in various cells and tissues, such as hepatocytes, thrombocytes, fibroblast, osteoclasts, neurons, brain, and many others, that aluminofluoride complexes might impair the calcium homeostasis [32]. The mechanism of the effect is not fully known at this point, but it seems that the changes in cytosolic $Ca^{2+}$ level are mostly due to the alterations of the phosphoinositide signaling system.

#### 3.3.1
#### *Alterations in Cytosolic $Ca^{2+}$ Level in AD*

The calcium hypothesis of brain aging and AD states that long-term, slightly elevated cytosolic $Ca^{2+}$ levels and/or disturbances in $Ca^{2+}$ homeostasis are the cellular mechanisms underlying neuronal aging [60, 61]. However, the precise details on the processes, which may lead to destabilization of calcium homeostasis in aging brain, are not known. Although many experimental data show the alteration of various $Ca^{2+}$ regulatory mechanisms during aging and

AD, the heterogeneity of types of neuronal preparations and experimental techniques does not allow full confirmation of the $Ca^{2+}$ hypothesis [62, 63].

Alteration in calcium homeostasis could affect most of the AD-related deficiencies. On the other hand, it is not clear whether changes in cytosolic calcium concentration ($[Ca^{2+}]_i$) are the result or the cause of pathogenic effects. Most of evidence regarding the changes of calcium homeostasis in aged neurons demonstrates an increase in resting cytosolic $Ca^{2+}$, and, after stimulation, a prolongation of $Ca^{2+}$ recovery towards the resting concentration [63].

It is difficult to get brain cells of AD patients while they are alive for laboratory studies. Therefore, various peripheral cells are used for parallel investigations of alterations in brain cells. The biochemical approach using peripheral cells is based on the assumption that the pathological disorder may lurk at the level of cell physiology and may be common for some peripheral cells and neurons. Measurements of $[Ca^{2+}]_i$ in peripheral human cells during aging and from AD patients bring controversial results. No age-related alterations in $[Ca^{2+}]_i$ were reported in human lymphocytes, granulocytes, and neutrophils, while basal $[Ca^{2+}]_i$ was increased in lymphocytes of AD patients [64]. Increased superficial binding of $Ca^{2+}$ and lower $[Ca^{2+}]_i$ were found in fibroblasts from aged and AD donors when compared to cells from young donors. Comparison of the activation-induced rise in $[Ca^{2+}]_i$ revealed parallel age-related changes in the different cell-types investigated. In agreement with these findings, the abnormality of various intracellular stores in cultured fibroblasts from AD patients was reported [65]. It is evident that a number of alterations in calcium homeostasis have been found in peripheral cells of patients with AD.

All these above mentioned controversial findings could not solely be attributed to methodological variations among different laboratories, as age-matched controls and AD patients were examined in the same conditions. However, the AD patients undergoing examinations are in different stages of the disease and, unfortunately, they are not well defined in some studies. Ripova et al. [64] measured the $[Ca^{2+}]_i$ in platelets isolated from the blood of AD patients in early stages in comparison with the groups of healthy young and age-matched controls and found a significantly decreased $[Ca^{2+}]_i$ in platelets of patients with AD. Our data also suggest that the buffering capacity is disturbed in platelets of patients with AD. Any stimulation of extracellular $Ca^{2+}$ influx results in increased cytosolic $Ca^{2+}$. Our experimental data show that platelets from AD patients have impaired compensatory mechanisms, resulting in movement of $[Ca^{2+}]_i$ towards the harmful levels. Such a disturbance was not observed in platelets from healthy age-matched donors [64]. We could therefore suggest that the alterations in cytosolic calcium regulatory mechanisms are not due to aging but might rather be connected with some pathogenic change such as the presence of the $\beta$-amyloid protein. In fact, one of the postulated mechanisms of the $\beta$-amyloid toxicity seems to involve a calcium disturbance accompanied by enhanced vulnerability to excitoxic stimuli. It has been recently demonstrated that the $\beta$-amyloid proteins form $Ca^{2+}$ channels across cell membranes [66]. We propose the possibility that AD might be associated with a generalized systemic defect in calcium homeostasis, which is not limited only to the brain neurons. Calcium overload may be detrimental to neurons, resulting in degeneration and death of brain cells in the late stages of AD.

### 3.3.2
### *The Effect of $AlF_x$ on Calcium Homeostasis*

Aluminum in micromolar concentrations was found to inhibit calcium pumping in endoplasmic reticulum. The $Ca^{2+}$ ATPase activity of rat brain and cerebellum was remarkably reduced and mitochondria showed increased $Ca^{2+}$ release in the presence of exactly estimated 50 μmol $L^{-1}$ $Al^{3+}$ [67]. Aluminum was found to be an important disrupter of intracellular calcium homeostasis, interfering also with the mitochondrial $Ca^{2+}$ pump, as well as activating an $Na^+$-$K^+$ ATPase – the antiport mechanism of ion exchange in the plasma membrane, which regulates the $Ca^{2+}$-$Na^+$ antiporter exchange [67].

We observed that the incubation of platelets or red blood cells in the presence of $AlF_x$ evoked changes in the phosphoinositide signaling system and a significant decrease of platelet $[Ca^{2+}]_i$ [68, 64]. This response was not different in platelets isolated from the blood of AD patients as compared with healthy age-matched subjects or young healthy subjects.

Chen and Penington [69] tested two hypotheses concerning the actions of $AlF_4{}^{1-}$ in the inhibitory effect of G-proteins on calcium channel activity. It has been mentioned above that $Ca^{2+}$ channels are also regulated by G-protein [31]. Earlier studies using intracellular fluoride, presumably made up in glass containing $Al^{3+}$, found that high-threshold L-type $Ca^{2+}$ currents are eliminated by this intracellular solution. L-type current components show very slow inactivation. Consequently, inhibition of a large L-type component of $Ca^{2+}$ current would increase the relative amplitude of a rapidly inactivating component when using a fluoride-containing recording solution. Chen and Penington [69] suggested that there can be a competition between the receptor and $AlF_4{}^{1-}$-stimulated G-protein activity, resulting in the short-term removal of tonic G-protein stimulation by $AlF_4{}^{1-}$ in the dorsal raphe neurons. Serotonergic dorsal raphe neurons have a small L-type $Ca^{2+}$ current and its abolition by fluoride is not very apparent. They also occasionally observed an increase in the rate of voltage-dependent $Ca^{2+}$ current inactivation. They investigated whether the interaction occurs at the level of the G-protein, or at the level of the interaction of the Gα-subunit with $Ca^{2+}$ channels. The main findings of these very extensive study were that intracellular $AlF_4{}^{1-}$ caused approximately one-third of the maximum tonic stimulation of the G-protein coupled to $Ca^{2+}$ channels of dorsal raphe neurons, consistent with a $G \cdot GDP \cdot AlF_4{}^{1-}$ complex resulting in a mimicry of the $G \cdot GTP$ complex (G-release).

### 3.4
### The Effect of $AlF_x$ on Changes of Cytoskeletal Proteins

AD is commonly characterized as the neurodegenerative disorder with the occurrence of senile plaques and neurofibrillary tangles. Neurofibrillar protein aggregates in patient brains contain abnormally phosphorylated protein $\tau$ [70]. Since phosphorylation is the regulatory mechanism of cytoskeletal protein assembly and function, it can be suggested that aluminofluoride complexes might induce important pathophysiological changes. The $\tau$ protein regulate the dynamics of the microtubule network, especially that involved in the axonal transport and neuronal plasticity. Tau belongs to a family of developmentally regulated isoforms generated by alternative splicing and phosphorylation. This generates several $\tau$ variants that interact with tubulin and other proteins. Therefore, $\tau$ might be influenced by many physiological regulations. In AD all six adult isoforms of $\tau$ can become maximally phosphorylated and can, rather than binding to microtubules, bind to each other, destabilizing the neuronal cytoskeleton [70].

The cytoskeletal network is responsible for the mechanical properties of the cell that modulate functions such as cell shape, locomotion, cytokinesis, and translocation of organelles. Experimental evidence suggests that the cytoskeleton also provides connections between cellular structures and presents a large surface area for interactions of various proteins and signaling molecules. Modulation of the cytoskeletal network may influence cell signaling, ion channels and intracellular calcium levels. Cytoskeleton is thus essential for regulation of cellular functions, cell integrity, and viability.

Microtubules of all eukaryotic cells are unbranched cylinders built up from the microtubule protein tubulin that occurs in two related forms ($\alpha$- and $\beta$-tubulin). Microtubule functions are based on their capacity to polymerize and to depolymerize. Tubulin is a GTP-binding protein. The binding of GTP is required for microtubule polymerization, whereas the hydrolysis of GTP is required for microtubule depolymerization. It has been reported that $AlF_4{}^{1-}$ inhibits tubulin depolymerization [16, 17, 71]. $AlF_x$ seems able to mimic the terminal phosphate only after its dissociation from the nucleotide but before its release from the protein. The aluminofluoride complex can bind to the tubulin molecule in GDP $\beta$-phosphate and mimic the GDP+P intermediate state. Their binding affinity is three-fold higher than that of phosphate. $AlF_x$ may in this way impair the polymerization-depolymerization cycle of tubulin. The effects of $Ca^{2+}$, 1,4,5-$IP_3$, and $AlF_x$ on the organization of spectrin in human red blood cell membrane skeletons were studied in the past in our laboratory [68]. We demonstrated that the phosphoinositide signaling system plays an important role in the shape maintenance and organization of the cytoskeletal network. The formation of holes, disorganization of the spectrin network, and shape changes in RBC were seen after 5 min of incubation with $AlF_x$. Here we show the changes in tubulin structure after 5 min incubation with $AlF_x$ (Fig. 3).

The breakdown or disorganization of the membrane skeleton can be followed by the release of some membrane-bound enzymes, such as AChE [43] or amyloid precursor protein (APP), which are crucial for the pathologic changes in the brains of AD patients. $AlF_x$ might affect the structure and function of cytoskeletal proteins by several routes [68]. It can activate various G-proteins and protein kinases, act as the analogue of GTP in the assembly-disassembly cycle, and affect the binding of cytoskeletal proteins to

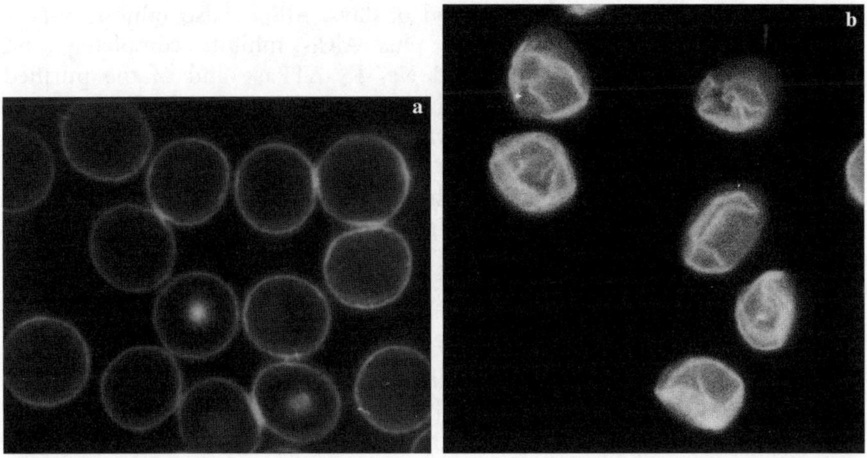

**Fig. 3.** Visualization of tubulin structures was carried out by indirect fluorescence, using monoclonal antibody TU-01 against $\alpha$-subunit of tubulin. The visualization of tubulin was done in cooperation with Prof. Dr. J. Palecek from the Faculty of Science, Charles University, Prague. By using 0.25% glutaraldehyde-1% Triton-X-100 fixation, we were able to detect the presence of tubulin as a diffusable fluorescence staining with a ring underlies the RBC plasma membrane (Fig. 3, **a**). The effect of 1 mmol $L^{-1}M$ NaF and 20 μmol $L^{-1}$ $AlCl_3$ is shown in Fig 3, **b**. Tubulin aggregated in an irregular shape inside the cells. Neither NaF nor $AlCl_3$ evoked the tubulin response to the extent observed with their combined action

membrane. It can affect the interactions between various cytoskeletal proteins. It has been documented that $AlF_x$ might affect the intracellular transport of vesicles with APP [72]. The splitting of APP is regulated by phosphorylation. Activation of protein kinase C increases the splitting of APP [73].

## 3.5 The Effect of $AlF_x$ on ATPases

Lunardi et al. [74] reported the inhibition of mitochondrial ATP synthase in the presence of $AlF_4^{\,1-}$. This enzyme couples the hydrolysis of ATP or its synthesis from bound ADP and phosphate. The terminal phosphate undergoes a sequence of coordination changes through which it is bound or released. ATP generation in mitochondria requires the association of the $F_1$ subunit with $F_0$ transmembrane subunit transporting protons. The binding of ADP and $P_i$ in a catalytic site of $F_1$ triggers conformational changes, which lock both of them into the site and induce the formation of pyrophosphate bonds by eliminating a water molecule [17].

In the presence of ATP, isolated $F_1$ functions as an ATPase. The binding of $AlF_4^{\,1-}$ into the catalytic site inhibits the ATPase activity. This inhibition is not reversed by elution of fluoride from solution or by addition of strong chelators of aluminum. Lunardi et al. [74] observed that no significant release of the complex occurred over a period of days. $AlF_4^{1-}$ also inhibits P-type cation-transport ATPases [75]. NaF plus $AlCl_3$ inhibits completely and reversibly the activity of the purified $Na^+$-$K^+$-ATPase and of the purified plasmalemmal $Ca^{2+}$-Mg-ATPase. An inhibition of the activity of the sarcoplasmic reticulum $Ca^{2+}$-Mg-ATPase was also observed. Due to the fact that $AlF_4^{1-}$ still inhibits ATPases, in the presence of GDP(S), and guanosine 5-[$\beta\gamma$-imido] triphosphate does not inhibit the ATPases, inhibition due to the activation of G-protein is unlikely. Aluminum alone at micromolar concentrations was found to inhibit $Ca^{2+}$-ATPase and $Na^+$-$K^+$-ATPase of rat brain and cerebellum [67].

The ATPase pathway thus must go through a pentacoordinated transition state for the $\gamma$-phosphate. The analogy with beryllium bonding that is strictly tetrahedral led Chabre to suggestion that $AlF_x$ cannot follow this route [17]. Wittinghofer describes in his article [21] an admirable methodological and intellectual approach, which led to further understanding of the mechanism of phosphoryl- transfer reactions using $AlF_x$ and proved that $AlF_x$ acts as the transition state analogue. The analogy with nitrogenase suggests a way in which ATP hydrolysis and a subsequent conformational change would break the interaction between the ATPase and the electron transfer system [21]. The structures of the residue that interact with $AlF_x$ have been determined for some enzymes and protein complexes, which contain a P-loop motif found in many ATP- and GTP-binding proteins.

The most important of the protein residues seems to be a lysine, which contacts a $\beta$-phosphate oxygen of ADP, and an aspartic acid, which is in a suitable location to serve as a potential general base for activation of water. An important element of the conformational change is a DxxGD motif, the glycine of which is involved in a main-chain interaction to a fluoride [21]. The second aspartic acid is homologous to the invariant glutamine in GTP-binding proteins. In the case of uridylate monophosphate kinase it was arginine 93, which interacted more strongly with one of the fluorides of $AlF_3$. Mutagenesis studies indicated that substitution of this arginine results in a 100- to 1000-fold reduced catalytic rate [18].

Structures of transition state analogue complexes of phosphoryl-transferring enzymes can be used to deduce whether the mechanism of the phosphoryl-transfer reaction is mostly dissociative, with a tetrahedral metaphosphate-like intermediate, or associative, with a pentavalent phosphorus (Scheme 2). Since the two mechanisms differ in charge distribution and bond order of the transition state, the location of positively charged amino acids and the distance of the $AlF_x$ to the leaving and attacking groups are crucial. The structures of transition state mimics show that the distances between aluminum and the leaving group and nucleophilic oxygen are intermediate between van der Waals and covalent bonds [18], suggesting a mostly associative character. According to Wittinghofer [21] the fact that, in most cases, positively charged groups contact the phosphoryl group to be transferred is an even stronger argument for an associative mechanism, since positive charges contacting the phosphoryl group would be anticatalytic in the case of a dissociative mechanism.

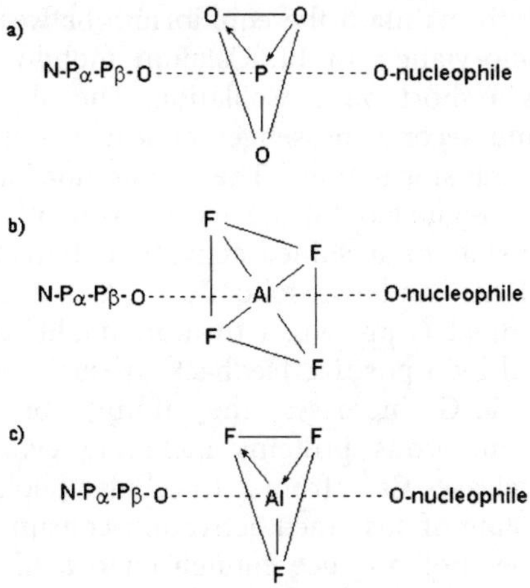


**Scheme 2.** Conformational changes of the $\gamma$-phosphate in **a)** phosphoryl-transfer reaction transition state K), and various species of $AlF_x$: **b)** $AlF_4^{1-}$, **c)** $AlCl_3$. Dotted lines indicate that the degree of bond making and bond breaking determines whether the transition is more dissociative, with a metaphosphate-like intermediate, or associative, with a pentavalent intermediate. Charges have been omitted for clarity. N = adenosine or guanosine. According to [16, 17, 21]

## 3.6
## Complexity in Signaling Systems on the Cell Level

Signaling via the large family of G-proteins can lead to many cellular responses ranging from regulation of intracellular messengers to gene transcription. Synergistic cross-talk interactions between G-coupled receptors have an important role in fine-tuning signals from multiple signaling pathways. At present we have no conceptual framework, which could allow us to integrate all knowledge about the regulatory mechanisms in processes of neurotransmission. Nevertheless, the widespread use of $Al^{3+}$ and fluoride salts in laboratory investigations has led to a jump in our understanding of the complexity of signal transduction processes.

We could approach the complex integrative view by the detailed study of the action of one small molecule, namely $AlF_x$. We could also judge its role in the etiology and pathogenesis of AD on the cell level. The knowledge from studies of the phosphoinositide signaling system and $Ca^{2+}$ homeostasis could help our understanding of the multifactorial nature of AD.

### 3.6.1
### *The State of PPI Signaling System Depends on the Physiological State of the Cell*

In our opinion, the final functional response after the generation of the first second messenger molecules of 1,4,5-$IP_3$ depends on the physiological state of the cell. The healthy cell maintains its content of ATP, $Ca^{2+}$ homeostasis, and keeps the ability to maintain the equilibrium between the breakdown of $PIP_2$ and the phosphorylation of PI. Calcium mobilization proceeds as a transient peak or as a short wave oscillation. The physiological functional response proceeds and second messenger molecules – 1,4,5-$IP_3$ and DAG – undergo metabolic transformation. The generation and degradation of second messengers is regulated by the mechanism of a negative feedback. $AlF_x$ may induce the state of sustained activation characterized by increased levels of 1,4,5-$IP_3$, DAG, and cytosolic $Ca^{2+}$ [32]. A disorder in calcium homeostasis or sustained G-protein activation might induce an amplification cascade regulated by a positive feedback. In such a case the breakdown of $PIP_2$ increases, DAG increases the affinity of protein kinase C, phosphorylation of numerous proteins including cytoskeletal proteins is enhanced, 1,4,5-$IP_3$ releases $Ca^{2+}$ from intracellular pools, $Ca^{2+}$ channels are opened, etc. Such a state of sustained activation consumes great amounts of energy. If the cell does not produce enough energy, the content of inositol lipids in the plasma membrane declines. At the same time, the level of cytosolic $Ca^{2+}$ is increased, because mechanisms of calcium homeostasis, such as the activity of the $Ca^{2+}$ ATPase, are disturbed. The generation of second messenger molecules is still regulated by positive feedback in such conditions. This may lead to disturbances of cell homeostatic mechanisms, cell damage, or apoptosis. Experimental observations support the view that $AlF_x$ may evoke the state of sustained activation with the stimulation of G-protein. The impairment of ATP synthesis can lead to energy depletion of the cell.

### 3.6.2
### *Interactions in Signaling Networks*

Moreover, the phosphoinositide signaling system interacts with many other transducing systems. The interactions between this system and the cAMP generating system have been well documented [27, 49, 76]. Different subtypes of cholinergic receptors also reveal multiple interactions. Many reactions proceed simultaneously or are overlapping. A biochemical link between tyrosine kinases and the phosphoinositide signaling pathway has been also suspected. It is possible that PI kinase may be a substrate for tyrosine kinase and, as well, that PLC may be regulated by tyrosine kinase-catalyzed phosphorylation. The release of arachidonate and subsequent synthesis of eicosanoids appear to be stimulated by the same receptors that control phosphoinositide hydrolysis. Experimental evidence suggests that protein kinase C may be involved in $PLA_2$ activation. There are multiple pathways involved in the phosphoinositide cascade that has been implicated in arachidonate release. On the other hand, arachidonate and eicosanoids may modulate PLC activity and the metabolic turnover of phosphoinositide. Prostaglandin $E_2$ and prostacyclin inhibit phosphoinositide breakdown and generation of inositol phosphates [76].

### 3.6.3
### *Complexity as Observed in the Experiments* in vitro

Shafer and coworkers [77] compared the effect of various concentrations of $AlCl_3$ and 5–30 mmol $L^{-1}$ NaF on muscarinic, adrenergic, and metabotropic receptor-stimulated phosphoinositide hydrolysis in hippocampal and cortical slices from rat brain. Their detailed study reveals the multiple possibilities of interventions into the phosphoinositide cascade and, simultaneously, the difficulties in the interpretations of results *in vitro*. In agreement with many others, Shafer and coworkers found that NaF stimulates inositol phosphate (IPs) accumulation as well as the cholinergic agonist carbachol, the adrenergic agonist norepinephrine, and the glutaminergic agonist quisqualate. The higher concentrations of $AlCl_3$ (0.5 mmol $L^{-1}$) inhibited IPs accumulation stimulated by agonists and that stimulated by NaF. They concluded that $AlF_4{}^{1-}$ is not responsible for the effects of $AlCl_3$ on receptor-stimulated IPs accumulation. There was a biphasic effect of $AlCl_3$ on agonist-stimulated IPs accumulation. The authors are aware that trace amounts of aluminum plus NaF stimulate the accumulation of IPs. The observed effects of the high concentrations of aluminum ions suggest their direct effect on G-proteins or PLC activity. However, the estimation of the total amount of IPs cannot reflect the hydrolysis of $PIP_2$, the major target of PLC after the receptor stimulation. The level of inositol 1-phosphate (1-IP) is about 30 times higher than that of 1,4,5-$IP_3$. The role of 1-IP in the cell physiology is not known, but some studies have suggested that changes in the level of inositol 1-phosphate in the brain are in parallel with excitation and inhibition [49]. It is very likely the increased $Ca^{2+}$ cytosolic level leads to the shift of phospholipase C activity to PI instead of $PIP_2$. Further studies of the effect of $AlF_x$ on the phospholipase C isoforms are therefore warranted.

The highly expert and detailed paper of Chen and Penington [69] could also demonstrate the complexity of the cellular system *in vitro*. $AlF_4{}^{1-}$ placed intracellulary in a patch pipette, activated a G-protein, resulting in a "tonic" inhibition of the $Ca^{2+}$ current of isolated serotonergic neurons of the rat dorsal raphe nucleus. Serotonin also inhibits the $Ca^{2+}$ current of these cells. After external bath application to and quick removal of serotonin from an $AlF_4{}^{1-}$ containing cell, there was a reversal or transient disinhibition of the inhibitory effect of $AlF_4{}^{1-}$ on the $Ca^{2+}$ current. A short predepolarization of the membrane potential to +70 mV, a condition that is known to reverse G-protein-mediated inhibition, reversed the inhibitory effect of $AlF_4{}^{1-}$ on the $Ca^{2+}$ current and brought the $Ca^{2+}$ current to the same level as that seen at the peak of the transient disinhibition current. With $AlF_4{}^{1-}$ in the pipette, the transient disinhibition phenomenon could be eliminated by lowering the pipette MgATP, or by totally chelating $Al^{3+}$ in the pipette. In the presence of $AlF_4{}^{1-}$, but with either lowered MgATP or extreme efforts to eliminate pipette $Al^{3+}$, the rate of recovery from serotonin on wash was slowed, a condition opposite to that where a transient disinhibition occurred. The putative complex of $AlF_4{}^{1-}$-bound G-protein appeared to free G$\alpha$-subunits, mimicking the effect on $Ca^{2+}$ channels of the G$\cdot$GTP complex. The ON-rate of the inhibition of $Ca^{2+}$ current, after a depolarizing pulse, by α-subunits released by $AlF_4^{1-}$ in the pipette was significantly slower than that of the agonist-activated G-protein. The OFF-rate of the $AlF_4^{1-}$-mediated inhibition in response to a depolarizing pulse, a measure of the affinity of the free G α-subunit for the $Ca^{2+}$ channel, was slightly slower than that of the agonist stimulated G-protein. In summary, agonist application temporarily reversed the effects of $AlF_4^{1-}$, making it a complementary tool to GTP-γ-S for the study of G-protein interactions. Based on the literature, Chen and Penington suggest a fairly parsimonious explanation of the sequence of events occurring after agonist application to a dorsal raphe cell containing $AlF_4^{1-}$. They propose that after several applications of serotonin, some G-proteins are in the basal state and some are activated by $AlF_4^{1-}$. This should produce only partial channel inhibition, allowing further inhibition by serotonin. This example is a large simplification of experiments and discussion in the above-mentioned study.

### 3.6.4 *The Unifying Hypothesis?*

A new unifying hypothesis of AD, which explains the mechanism whereby aluminum could induce, amplify, and orchestrate a cascade of neurotoxic events characteristic of AD, has been proposed by Exley [5, 10]. He has identified a mechanism through which a truly chronic exposure to aluminum would bring about subtle and persistent changes in neurotransmission, which, in time, could instigate the cascade of events known collectively as AD. This mechanism involves the potentiation of the activities of neurotransmitters by the action of aluminum-ATP at ATP receptors in the brain. There is growing evidence that ATP plays an important role not only as the universal donor of energy but also as an important multi-functional signaling molecule. In the brain, ATP acts upon ionotropic (P2X) and metabotropic (P2Y) receptors to optimize the activities of neurotransmitters including acetylcholine, glutamate, and γ-aminobutyric acid. Two characteristics make ATP a particularly interesting neurotransmitter. After release, ATP is rapidly degraded to adenosine, which can also act as a neurotransmitter and modulate the transmitter release at presynaptic terminals. Also, ATP ionotropic receptors are highly permeable to calcium ions and, unlike NMDA receptors, are also functional at negative potentials. Therefore, ATP receptors may represent a new route for synaptically gated calcium entry into dorsal horn neurons at the cell resting potential and could be involved in the phenomena of synaptic plasticity and central sensitization, and in modulating synaptic transmission under particular conditions, such as during inflammation. The extracellular action of ATP is mediated via release of intracellular ATP. In the brain, ATP is secreted by neurons, glia, and endothelial cells. The changes in the extracellular concentration of ATP could thus evoke various and multiple cascades of events. Of great importance in the Exley hypothesis is the fact that a very subtle change in the extracellular ATP concentration could evoke a very complex network of events in the brain. Aluminum is bound by ATP to form a complex, which is orders of magnitude more stable than that with magnesium

[5]. Exley elaborated the complex theory by describing how a physiologically significant concentration of aluminum could induce various pathophysiological changes. According to our hypothesis, it seems that fluoride could further potentiate the action of aluminum. $AlF_x$ evokes the effects via G-proteins probably at very low concentrations, even under such situations when the actual content of aluminum in the brain is not increased. It seems that scientists still need more significant data to be sure about the causal relationship between the action of a small ions and molecules, such as $Al^{3+}$ and $AlF_x$, and disturbances of human health. Science has already accumulated evidence about their detrimental effects on the whole organism.

## 4 The Holistic Complexity

AD is probably an example of a multifactorial disease whose pathophysiology cannot be understood on the basis of one altered molecule or gene. The interpretation of laboratory investigations using isolated animal and human cells or tissues on the intact human organism could be discussed. However, many ecological and clinical studies bring evidence about the detrimental effects of the synergistic action of aluminum and fluoride ions in humans. Several processes are implicated in the uptake and distribution of aluminum and fluoride in tissues and organs of the human body. The severity and the development of the symptoms depend on a person's age, genetic background, nutrition status, kidney function, and many other factors. It is evident that the dietary intake of aluminum and fluoride varies all over the world. It is not surprising that various ecological studies from different geographical locations might bring controversial or less significant results. In epidemiological studies many additional limitations and variables may occur. The history of the use of fluoride in the prevention of caries provides the example that the evaluation and interpretation of obtained evidence is, to some extent, subjective. Individual susceptibility to AD will be as relevant as the truly chronic exposure to an increased load of aluminum and fluoride from the environment and food chain. It might be difficult to decide whether aluminum and fluoride are the main responsible factors for reported risks or benefits. How many subjects shall we need to prove their hidden danger?

### 4.1 Aluminum and Fluoride in the Environment

It is generally believed that exposure to aluminum is mainly through oral intake, and the major sources are drinking water, residues in foods, cooking utensils, food, and beverage packaging. There are very few data on the oral bioavailability of aluminum from food. An excellent comprehensive review about oral bioavailability of aluminum from drinking water and food has been published recently by Yokel [8]. According to his conclusion, the major source of aluminum absorption daily is food, not water. Exposure is also possible from aluminum-containing medications (e.g., antacids and buffered aspirins), parenteral solutions, and vaccinations containing aluminum adjuvant. The daily intake of aluminum from medications can greatly exceed that from the diet and water. Occupational exposure to aluminum occurs in the refining of the primary metal and in secondary industries that use aluminum products.

Fluorine was not produced by industry until the 2nd half of the 20th century and its amount in the water sources and food was negligible. The development of fluorine chemistry was followed by its increasing consumption in military, pharmaceutical, and electronic industry. During the last 50 years, fluorides have been widely used in tooth decay prevention. Thanks to the reduction of dental caries observed both in children and adults due to the increased fluoride intake, fluoride supplementation of different food sources, oral cosmetics, and prophylactic dental devices has begun. Fluoridation of water supplies and industrial fertilizers increase the amount of this element in agricultural products and food chains. Fluoride comes from fluoridated water, from crops grown with fluoridated water, from medicines, dental products, pesticides, fertilizers, and fuels [78]. It may thus be justifiably postulated that the fluoride intake may be in some individuals supraoptimal and, together with the increased intake of aluminum may exceed their tolerable limits. The contents of fluoride salts in drinks and food (www.bruha.com/fluoride) contribute significantly to the total dietary intake and exceed the recommended daily dose (2 mg per day).

The quantity of fluoride and aluminum in living environment tends to increase, causing their increase in human food chains and resulting in their increased load to the human body. Several studies reported a high content of fluoride and aluminum in all tested Chinese, Indian, and herbal teas [79, 80]. Tea is the second most-consumed beverage after water in the world. That the aluminum present in tea is indeed resorbed in the simultaneous presence of fluoride was demonstrated in healthy male volunteers after drinking equal volumes (1.2 L) of tea, coffee, or tap water on separate days. In every case the amount of the urinary excreted aluminum increased on the day when tea was taken [81]. The results indicated that tea consumption must be considered in any assessment of the total dietary intake of aluminum in human beings. A significant fact to be taken into account is that the increasing fluoride and aluminum load may be cumulative and health threatening. The acid rains, fluoridation of water supplies, the use of fluoride and aluminum in industry as well as medicine started the era of supplementation of human bodies with these ions as never before in the history of the human race.

## 4.2
## Aluminum and Fluoride in the Human Blood and Body Fluids

The small $Al^{3+}$ cation with a high charge does not exist in the free form under physiological conditions. In mildly acid intracellular conditions, aluminum is bound to phosphate. Aluminum forms soluble salts with some acids, for example, with citric acid [82], or complexes with some compounds, for example maltol [80]. It seems that the oral bioavailability of aluminum is very low, about 0.3% [8]. The oral aluminum bioavailability was not found to be significantly different between young and aged subjects [83]. The proximal intestine appears to be the primary site of aluminum absorption. The putative mechanisms of absorption include diffusion, carrier- and vesicular-mediated transport as well as paracellular diffusion between the mucosal cells in the gastrointestinal tract.

Calcium channels are suggested to mediate intestinal aluminum absorption [84]. Aluminum distributes unequally to all tissues. The values given for the aluminum levels in the human body vary widely. Under normal conditions the aluminum concentration in blood is 5–10 μmol $L^{-1}$ [14, 85]. The mean aluminum concentration in human milk was 23.4 ± 9.6 μg $L^{-1}$ and did not differ significantly between colostrum, intermediate-stage, and mature-stage milk [86].

The estimation of fluoride concentration in human body fluids has been widely performed. However, data are expressed in various units. Kissa [87] reported that the mean plasma fluoride level in healthy subjects is between 20 to 60 μg $L^{-1}$ (1–3 μmol $L^{-1}$). Others [88, 89] present L-1 plasma fluoride concentrations in the range of 38.8–88.6 mol. The average fluoride concentration in plasma of 8–16-year-old students from Winterthur (Switzerland) was 12.7 ± 3.8 μg $L^{-1}$ (0.65 μmol $L^{-1}$) [90]. Human breast milk (samples were collected from 57 lactating mothers) contains 0.019 ± 0.004 ppm of fluoride [91].

The accepted therapeutic range of the serum fluoride levels has been suggested to be 5–10 μmol $L^{-1}$ and the "therapeutic window" 95–100 μg $L^{-1}$ (5 μmol $L^{-1}$) is not assumed to be overtly cytotoxic [92].

The normal concentration of fluoride in saliva is about 1 μmol $L^{-1}$, which is somewhat less than that in plasma. The concentration of fluoride in whole saliva mirrored the fluoride concentration in plasma, but at a lower level [93].

The Swiss study [90] reports the urinary fluoride concentrations estimated in 8–16-year-old students from various areas. The fluoride concentrations in urine were higher in Basel and Davos (0.62 ± 0.35 mg $L^{-1}$; 0.61 ± 0.42 mg $L^{-1}$) (32 μmol $L^{-1}$) than in Berne and Winterthur (0.46 ± 0.42 mg $L^{-1}$; 0.50 ± 0.31 mg $L^{-1}$) (24 μmol $L^{-1}$). A relatively high natural fluoride content (0.3 mg $L^{-1}$ to 16 μmol $L^{-1}$) in the drinking water explained the difference in urine fluoride concentration between students from Davos and Berne + Winterthur with domestic salt fluoridation.

The increased serum and urinary fluoride concentrations of workers exposed to hydrofluoric acid in the working environment were reported [94]. There was a linear relationship between mean serum and urinary fluoride concentrations and hydrofluoric concentration in the workplace. Mean fluoride concentrations of 82.3 μg $L^{-1}$ (4.5 μmol $L^{-1}$ in serum and 4 mg $L^{-1}$ (210 μmol $L^{-1}$) in urine were found at an atmospheric HF concentration of 3 ppm.

The individual fluoride exposure and the corresponding body fluid levels were studied in 41 workers in an aluminum plant in Sweden. During the shift (8 h) personal air samplings were performed, plasma fluoride levels and urine fluoride excretion determined. The average total fluoride exposure was

0.91 mg $m^{-3}$ of which 34% was gaseous fluoride. The mean fluoride plasma level before the shift was 23 μg $L^{-1}$ (1.2 μmol $L^{-1}$) and increased on average to 48 ng/ml (range 14–151 μg $L^{-1}$) at the end of the shift [95]. There was a high correlation between fluoride renal clearance and urinary flow. It seems therefore that a high fluid intake during the shift increases the capacity of the kidney to excrete fluoride and decrease the levels of fluoride in the body.

The increase of fluoride concentration in the blood has been reported in hemodialysed patients. Arnow et al. [96] reported that the serum concentrations of fluoride in the sick patients were markedly increased to as high as 716 μmol $L^{-1}$.

## 4.3
## The Action of Aluminum Ions and Fluoride in the Whole Organism

It is known that aluminum binds to fluoride in the diet [97]. The natural barrier systems and various physiological ligands for aluminum were efficient buffers preventing the increased intake of this metal in natural conditions. Most aluminum ingested leaves the intestines without resorption and only a small percentage of it reaches the blood stream [8, 98]. In the blood, aluminum is bound to transferrin and to albumin [8] and transported for renal elimination. Transport of aluminum from extracellular fluid to blood is believed to be mediated by a monocarboxylate transporter [5]. It is implied that the transport mechanism uses citrate as its vehicle. Measurements of uptake of radioactive soluble aluminum ($^{26}Al$) from stomachs of rats showed that trace amounts of $^{26}Al$ directly entered the brain [99]. Despite the extensive use of aluminofluoride complexes in experiments *in vitro*, the mechanism of their transport across the plasma membrane was not studied in detail. However, their effect on numerous intracellular reactions and functions was observed. Aluminum has been found in various human tissues in concentrations between 1–6 μg $g^{-1}$ dry weight [85]. Higher concentrations were found in bone and skin. Concentrations almost 10 times higher were seen in the lungs, probably due to deposition of particles from inhaled atmospheric dust [8, 85].

### 4.3.1
### *Endothelial Cells and Blood Vessels*

Endothelial cells and the kidneys play probably an important role in the long-term homeostasis as well as in the development of AD. Endothelial cells synthesize and release vasorelaxant and vasoconstrictor substances. The vasoconstrictor mediators, endothelins, were isolated, purified, sequenced, and cloned [100]. The human endothelin receptor has seven membrane helices and is apparently G-protein-coupled. Endothelin-induced smooth muscle contraction involves the following processes: activation of PLC, $PIP_2$ hydrolysis, generation of 1,4,5-$IP_3$, accumulation of DAG, mobilization of intracellular $Ca^{2+}$ with facilitation of $Ca^{2+}$ influx, activation of PKC, activation of $PLA_2$ and arachidonic acid release, activation of PLD, and stimulation of $Na^+$-$H^+$

exchange. It can be therefore suggested that $AlF_x$ would influence markedly the action of endothelins.

Endothelial cells produce prostacyclins in response to ATP. $AlF_4{}^{1-}$ enhanced the stimulatory effect of ATP on prostacyclin release [101]. A dose- and time-dependent generation of inositol phosphates, release of arachidonic acid, and the production of prostacyclin ($PGI_2$) in human umbilical vein endothelial cells was described [102]. Stimulation of arachidonic acid release and prostacyclin production by $AlF_4{}^{1-}$ was described in cultured endothelial cells from rabbit coronary microvessels [103].

ATP stimulates the synthesis of platelet-activating factor, which has an important function in mediating the adherence of inflammatory cells to the endothelium. The addition of $AlF_4{}^{1-}$ to endothelial cells resulted in the production of platelet-activating factor with a maximal effect within 20–60 min of exposure. $AlF_4{}^{1-}$ also augmented the production of platelet-activating factor, which occurs in response to a hormonal agonist [104]. In addition, submaximal concentrations of $AlF_4{}^{1-}$ converted an ineffective hormonal agonist to a maximally effective agonist.

Boonen and De Mey [105] investigated G-proteins' involvement in the resistance of arterial smooth muscle to contractile agonists. $AlF_4{}^{1-}$ induced contractions dependent on the presence of $Ca^{2+}$. Increased $PIP_2$ breakdown and the generation of inositol phosphates after the action of $AlF_4{}^{1-}$ was observed [106] in permeabilized vascular smooth muscle.

NaF evoked contractions by stimulating L-type $Ca^{2+}$ channels in rabbit femoral arteries, while $AlF_4{}^{1-}$ stimulated PLC, which produced additional muscle activation [107].

The modern view of the mammalian blood-brain barrier (BBB) has shifted from a purely anatomic concept to a more physiological and dynamic definition. Morphologically, a lining of specialized endothelial cells forms the BBB. Specialized regions of intercellular contact (tight junctions) that prevent leakage of blood-borne substances into the brain parenchyma characterize the cells of the intraluminal portion of brain capillaries. The failure of the structural integrity and function of the BBB plays a pivotal role in the pathogenesis of many diseases of the central nervous system. The presence of fluoride caused more aluminum to cross the blood-brain barrier and be deposited in the brain of rat [108]. Varner and coworkers studied alterations in the nervous system resulting from chronic administration of $AlF_3$ or equivalent levels of fluoride in the form of sodium fluoride. Twenty-seven adult male Long-Evans rats were administered one of three treatments for 52 weeks. The aluminum levels in samples of brain and kidney were higher in both the $AlF_3$ and NaF groups relative to controls. The effects of the two treatments on cerebrovascular and neuronal integrity were qualitatively and quantitatively different. Aluminum-induced neural degeneration in rats is greatly enhanced when the animals were fed low doses of fluoride. Animals develop both symptoms and brain lesions that are similar to those found in AD [108].

The kidney is both a source of endothelin generation and an important target organ of this peptide.

### 4.3.2 *The Kidney*

The effects of aluminofluoride complexes on the kidney were studied using glomerular mesangial cells, proximal tubular cells, and inner medullar collecting tubule cells of rat kidney.

Bradykinin elicits a complex response in the renal glomerulus, including reduction in the glomerular capillary ultrafiltration coefficient. The effects of bradykinin and vasopressin on calcium mobilization in renal glomerular mesangial cells of rat were studied using the calcium-sensitive fluorescent probe Indo 1 [109]. Both hormones were found to cause a transient concentration-dependent rise in the cytosolic $Ca^{2+}$ concentration followed by a sustained secondary rise. $AlF_4{}^{1-}$ induced a transient rise of cytosolic $Ca^{2+}$, which was abrogated by prior exposure of the cells to pertussis toxin. In contrast, the responses to bradykinin and vasopressin were unaffected by pertussis toxin.

The proximal tubule is responsible for the reabsorption of the major part of the glomerular ultrafiltrate. Proximal cells possess a brush border on their luminal surface. This membrane contains intrinsic transporting proteins like ion channels and ionic pumps. Hormones regulate the transport of ions and the involvement of adenylate cyclase has been well documented. $AlF_4{}^{1-}$ was the most effective stimulator of adenylate cyclase. A 9.3-fold peak increase in activity was observed in the basolateral membranes after its action. On the other hand, brush border membranes showed only a small increase [110]. The activity of amiloride-sensitive $Na^+$-$H^+$ exchanger is regulated by cAMP-dependent protein kinase. Incubation of vesicles, as prepared from renal brush border membranes of rat in the presence of 10 mmol $L^{-1}$ NaF and 10 μmol $L^{-1}$ $AlCl_3$, resulted in a significant inhibition of amiloride-sensitive $^{22}Na$ uptake [111]. Incorporation of GTP(S) also resulted in significantly reduced $^{22}Na$ uptake. cAMP-dependent protein kinase activity, strongly stimulated by exogenously added cAMP in these vesicle preparations, was not stimulated by GTP(S). These findings suggest that the amiloride-sensitive $Na^+$-$H^+$ exchanger could be regulated by G-proteins, independently of adenylate cyclase and cAMP-dependent protein kinase.

These data document that the ion transporting processes are affected by $AlF_4{}^{1-}$ in kidney tubular cells. $AlF_4{}^{1-}$ stimulates adenylate cyclase, inhibits amiloride-sensitive $Na^+$-$H^+$ exchange regulated by cAMP-dependent protein kinase, enhances epidermal growth factor-stimulated prostaglandin production, and mimics vasopressin and bradykinin induced $Ca^{2+}$ mobilization. It is suggested that $AlF_4{}^{1-}$ can affect the activity of many other ion channels and enzymes in the kidney. Borke and Whitford [112] examined the effect of chronic fluoride ingestion on ATP-dependent $^{45}Ca$ uptake by rat kidney membrane vesicles to characterize the mechanism by which fluoride alters calcium transport in the kidney. Twenty weaning female Sprague-Dawley rats were raised on low-fluoride (0.9 mg $L^{-1}$), semi-purified diet with a $Ca^{2+}$ concentration of 400 mg/100 g diet. Rats were divided into four groups and were fed ad libitum deionized water containing fluoride at concentrations of

0, 10, 50, or 150 mg $L^{-1}$ added as NaF for six weeks. This consumption produced plasma fluoride levels of <0.4, 2, 7, or 35 μmol $L^{-1}$, respectively. ATP-dependent $^{45}Ca$ uptake was significantly lower in the 150 mg $FL^{-1}$ exposure group than in the controls. Slot blot analysis of kidney homogenates with specific $Ca^{2+}$-pump antibodies showed significantly less endoplasmic reticulum $Ca^{2+}$-pump protein and plasma membrane $Ca^{2+}$-pump protein in all treatment groups than controls [112]. Both $Ca^{2+}$-pumps are transport molecules of great importance in the regulation of $Ca^{2+}$ homeostasis. The authors suggest that chronic, high fluoride ingestion producing high plasma fluoride levels may occur in humans and may affect calcium homeostasis.

### 4.3.3 *The Endocrine Glands*

The endocrine glands such as the thyroid, the pituitary gland, and the pineal gland, are extremely sensitive to fluoride and accumulate aluminum.

#### *4.3.3.1* *Thyroid*

Fluoride effects on the thyroid can be observed on many different levels. Of particular importance relating to G-protein activation is the ability of $AlF_x$ to clone the role of thyroid-stimulating hormone (TSH). Allgeier with coworkers [113] demonstrated that TSH led to the activation of two forms of Gs (Gs short and Gs long) as well as of Gq and G11, demonstrating that signaling pathways induced by TSH already bifurcate in the course of the receptor-G-protein interaction and concluded that TSH activation is mediated by Gq/11 and Gs, respectively.

Fluoride is used in laboratory animals specifically to substitute for TSH. There is a direct dose-response relationship with iodine: the higher the fluoride intake the lower is the iodine in the system. The synergistic action of the thyroid on fluoride toxicity has been reported since 1940. The major areas of iodine deficiency are identical to endemic fluorosis areas. The comparison of fluoride toxicity symptoms and symptoms of thyroid disorders has been reviewed by Schuld (www.bruha.com/fluoride/html/symptoms-hypo-f.htm). The functional changes of the hypophysis-thyroid gland system caused by disorders of the regulatory chain and fluorine impact on thyroid hormones metabolism at the level of target cells were also reported. Regarding the crucial role of the thyroid in regulation of growth, development, and metabolism of many tissues, $AlF_x$ might influence the proper function of the entire human body, including mental abilities.

#### *4.3.3.2* *Pineal Gland and Melatonin*

Knowledge of the pineal gland's function in animal and human organisms has been collected during the last thirty years. The experiments and observations point to an unbelievable number of functions regulated by the pineal gland. The physiological response depends on age, sex, physiological condition of the animal, time of day and season. The human pineal gland is a very small oval organ (it weighs 150 mg) connected to the surface of the 3rd brain ventricle. The pineal gland originates from the nerve tissue of the mesencephalon in the embryonic state. The brain and the pineal gland have a nerve and circulatory connection. The pineal gland is highly vascularized with no hematoencephalic barrier and has a very high perfusion rate of blood, second only to the kidney.

Melatonin is the chief pineal hormone. It is formed from its precursor, serotonin. The concentration of serotonin in the pineal gland is about 10 times higher than in other organs. Production of serotonin and melatonin is cyclical and is influenced by light. Production of serotonin increases during the day, while production of melatonin increases during the night. The pineal gland secretes melatonin into the blood and other body fluids. Measurable quantities of melatonin have been found in the cerebrospinal fluid, saliva, fluid of the eye, seminal fluid, breast milk, and in amniotic fluid. $AlF_x$ was used to indicate that the effects of melatonin are mediated by heterotrimeric G-protein [114].

From the evidence accumulated so far, it seems likely that the cAMP signal transduction pathway will be a major effector of a stimulatory signal to the pars tuberalis, which can be regulated by melatonin [115]. The effect of aluminum as $AlF_4{}^{1-}$ has been studied on inositol phosphate accumulation, calcium mobilization, and cyclic AMP production in ovine pars tuberalis cells [116]. In the presence of 10 mmol $L^{-1}$ LiCl, $AlF_4{}^{1-}$ stimulated the net accumulation of inositol phosphates over a 40-min incubation. Lithium is a known inhibitor of phosphatases in the inositol phosphate-recycling pathway. The results show the existence of a lithium-sensitive phosphoinositide signaling.

Consistent with these findings, $AlF_4{}^{1-}$ increased intracellular calcium; although this response was attenuated in calcium-depleted medium, indicating that the calcium response comprises both intracellular and extracellular components. $AlF_4{}^{1-}$ blocked the increase in cyclic AMP stimulation by 1 micromolar forskolin, being as effective as melatonin, achieving approximately 90% inhibition. The authors proposed that these results are consistent with the interpretation that $AlF_4{}^{1-}$ activates many G-protein-mediated responses, and thus imply that the inhibitory pathway for cyclic AMP predominates over the stimulatory arm.

The pineal gland produces various other hormones and mediators such as dopamine, norepinephrine, glutamic acid, $\gamma$-aminobutyric acid, and neuropeptides. Physiologists speculate that the activity of the pineal gland is regulated primarily by light signals: the pineal gland is considered to be a transducer of photoperiodical information. Light signals, which are registered in the eyes, go to the pineal gland via nerves of the suprachiasmatic nuclei of the hypothalamus. This structure functions as a biological clock. Suprachiasmatic nuclei in the hypothalamus are the oscillator, which gives biological rhythms and synchronizes the physiological activities of the organism. Nadakavukaren with coworkers [117] observed $AlF_4{}^{1-}$ stimulated accumulation of inositol phosphates within the suprachiasmatic region of the rat hypothalamus. Melatonin is responsible for regulating numerous life processes, including aging. The highest levels of melatonin are generated in young children. Luke [118] found very high levels of fluoride (9000 ppm) in the calcium hydroxyapatite crystals produced by the human pineal gland (she analyzed 11 cadavers). These levels are at, or higher than, the fluoride levels in the bones of people suffering from skeletal fluorosis. In animals (Mongolian gerbils) treated with fluoride, Luke found that melatonin production was decreased.

### 4.3.4
### *Aluminum- and Fluoride-Induced Apoptosis*

The programmed cell death – apoptosis – has been explained as a defective cell cycle resulting in cell death instead of cell division. Apoptosis is implicated in the pathology of many diseases. AD is connected with a progressive neuronal degeneration and death of cholinergic neurons. Some authors demonstrated that aluminum could induce the apoptotic degeneration of rat hippocampal neurons and astrocytes [119]. After 8–12 days exposure, aluminum (1 mmol $L^{-1}$) caused strong changes in the morphology of astrocytes including shrinkage of cell bodies and retraction of processes. Aluminum-induced degeneration of astrocytes involved the DNA fragmentation, characteristic of apoptosis. Astrocytes may regulate the expression or function of proteins in the neuronal cytoskeleton. The high concentration of aluminum used in these experiments may be rather unlikely to occur in vivo, since the concentrations of aluminum in human blood and tissues are usually very low. However, the mechanisms for the exclusion of aluminum from the human organism may be impaired during aging and the exposure to aluminum may be long-term. A higher amount of aluminum was found in the human brain with AD than in brains of age-matched healthy controls [2]. The presence of fluoride ions results in alterations in neuronal and cerebrovascular integrity and a greater level of aluminum in the kidney and brain [108].

Loweth et al. [120] showed that fluoride induces apoptosis in clonal pancreatic $\beta$ cells and in the cells of normal rat islets of Langerhans. The process may reflect the formation of $AlF_4{}^{1-}$ since it was inhibited by the aluminum chelator desferrioxamine. Recent studies provide evidence that apoptosis of pancreatic $\beta$ cells is important in the early etiology of diabetes mellitus. Treating thymus lobe cells with aluminofluoride complexes also provoked apoptosis of a wider range of thymocyte subtypes [121] with an accumulation of inositol phosphates. The responses to aluminofluoride complexes were not prevented by inhibitors of tyrosine kinases, suggesting that unidentified G-proteins which couple to phospholipase C activation may also be capable of initiating apoptosis by a route independent of the T cell receptor.

## 4.4
## Evidence about the Neurotoxic Effects of Aluminum plus Fluoride in Humans

Most of the ill effects caused by fluoride were first recognized among workers in aluminum factories, where fluoride and aluminum are present in high concentrations. Observation of industrial fluorosis and bronchial-related disorders in such workers have been reported since the 1930s. Psychiatric disturbances were also reported in aluminum smelter workers. The study of persons living near an enamel factory reports a distinct decline in mental activity, poorer memory, inability to coordinate thoughts, and reduced ability to write [122].

Elevated aluminum levels have been implicated as the cause of dialysis encephalopathy or dementia in renal failure patients undergoing long-term hemodialysis [85]. Some patients used aluminum-containing medications. Moreover, patients with renal failure cannot remove aluminum from the blood. Dialysis dementia can arise after three to seven years of hemodialysis treatment. Speech disorders precede dementia and convulsions. Since many hemodialysis units rely on systems to purify fluoridated tap water, it is likely that many patients are being exposed inadvertently to increased concentrations of fluoride and aluminum. Increased serum fluoride concentration and fluoride intoxication have been also observed in chronic hemodialysis patients. Arnow et al. [96] reported that 12 of 15 patients receiving dialysis treatment in one room became acutely ill, with multiple non-specific symptoms and fatal ventricular fibrillation. Death was associated with longer hemodialysis time and increased age compared with other patients who became ill.

Since fluoride has been used in the prevention of tooth decay for over 50 years, many studies reporting and evaluating the risks and adverse effects of fluoride on the human organism were published during the same period [78, 122, 123]. Reduction of children's intelligence, various psychiatric symptoms in adults, such as memory impairment, and difficulties with concentration and thinking were reported [78]. Aluminum, the most abundant metal of the earth's lithosphere, is everywhere. Vaccines, allergy skin tests, 25% human serum albumin, baby skin creams, baby diaper wipes, and antacids, which are frequently given to infants, are extremely high in aluminum. For adults, the accumulation continues from suntan lotion, cookware, and aluminum cans and skin moisturizers. Such items as deodorants, vaginal douches and baby wipe not only have high aluminum content, but also are applied to areas where there is a far greater tendency to absorption through the skin.

The synergistic action of aluminum ions with fluoride may be the underlying mechanism of the observed neurotoxic effects of fluoride. Chronic exposure of humans to $AlF_x$ begins in the fetus. Elevated fluoride content was found in embryonic brain tissues obtained from required abortions in areas where fluorosis was prevalent [78, 124]. These studies showed poor differentiation of brain nerve cells and delayed brain development. High fluoride exposure appears to weaken mental function among children, as well as adults [125, 126].

The increasing content of fluorides and aluminum in food chains has raised the possibility that the near future will supply us with much more data about the neurotoxic effects of aluminum plus fluoride on humans.

## 5 Conclusions

Any endeavor to think about the possible action of aluminofluoride complexes could provide a lesson which could draw us nearer to understanding the multiple nature of the etiology and pathogenesis of AD, providing a better chance of prevention of this most devastating brain disease. The search for AD prevention probably waits for a change in the concepts and strategy of scientific research from the reductionistic approach, which supplies us with the parts of the puzzle, to the integration of these parts into a multidimensional and non-linear whole. No one can predict what happens in the human body after a truly chronic exposure to an increased content of aluminum and fluoride in body fluids and in various tissues.

The aluminum concentration in human blood is comparable with that used in the laboratory investigations. The fluoride concentration in human blood is still many times lower than that used in the laboratory. It is difficult to predict the actual concentrations of aluminum and fluoride in intracellular compartments. But the increase of serum concentrations of fluoride in chronic hemodialysis patients to 756 $\mu$mol $L^{-1}$ resulted in illness or death.

Recently, we can witness many discussions about the benefits and risks of the fluoride supplementation. Also the question of aluminum toxicity in men has been discussed. The understanding of the mechanisms of their synergistic action could allow us to explain numerous observations about the effects of increased load of fluoride and aluminum in the environment and to reevaluate their widespread use. Understanding the role of phosphate and G-proteins in cell signaling forces us to accept the fact that aluminum in the environment, water, and food chains followed by fluoride ions could evoke various and multiple pathological symptoms.

G-proteins take part in an enormous variety of biological signaling systems, helping control almost all important life processes. The origins of many human diseases are in the functioning (and malfunctioning) of signaling components [25]. Pharmacologists estimate that up to 60% of all medicines used today exert their effects through G-protein signaling pathways [127]. However, it is evident that processes of signal transmission amplify a signal from a molecule giving the false information, namely $AlF_x$. In light of the published findings, the long-term action of $AlF_x$ represents a hidden but serious and powerful risk factor for the development of AD. The awareness of the health risks of increasing load of aluminum ions and fluoride as a new ecotoxicological phenomenon would contribute significantly to the decline of risks of intelligence decreases in children and adults, and many other severe disorders in the 21st century.

**Acknowledgement.** We gratefully acknowledge the support of the Grant Agency of Charles University Prague and of the Internal Grant Agency of Ministry of Health Czech Republic.

# 6 References


1. McDermott JR, Smith AI, Iqbal K, Wisniewski HM (1979) Neurology 29: 809
2. Galle P, Meyrignac C, Heine P (1984) Cr Acad Sci Ser III, Sci Vie 299: 535
3. Crapper MC, Lachlan DR (1986) Neurobiol Aging 7: 525
4. Zatta P, Giordano R, Corain B, Bombi GG (1988) Med Hypoth 26: 139
5. Exley C (1999) J Inorg Biochem 76: 133
6. Flaten TP (2001) Brain Res Bull 55(2): 187
7. Rondeau V, Commenges D, Jacqmin-Gadda H, Dartigues JF (2000) Am J Epidemiol 152: 59
8. Yokel RA, McNamara P (2001) Pharmacol Toxicol 88: 159
9. Savory J, Exley C, Forbes WF, Huang Y, Joshi JG, Kruck T, McLachlan DR, Wakayama I (1996) J Toxicol Environ Health 48: 615
10. Exley C (ed) (2001) Aluminium and Alzheimer's Disease. The Science that Describes the Link. Elsevier Science, New York
11. Saunders AM (2001) Pharmacogenomics 3: 239
12. Strunecka A, Patocka J (1999) Fluoride 32: 230
13. Rall TW, Sutherland EW (1958) J Biol Chem 232: 1065
14. Sternweis PC, Gilman AG (1982) Proc Natl Acad Sci USA 79: 4888
15. Goldstein G (1964) Anal Chem 36: 243
16. Bigay J, Deterre P, Pfister C, Chabre M (1987) EMBO J 6: 2907
17. Chabre M (1990) Trends Biochem Sci 15: 6
18. Schlichting I, Reinstein J (1999) Nat Struct Biol 8: 721
19. Martin RB (1996) Coord Chem Rev 149: 23
20. Antonny B, Chabre M (1992) J Biol Chem 267: 6710
21. Wittinghofer A (1997) Current Biology 7: R682
22. Oliveira L, Paiva ACM, Sander C, Vriend G (1994) TIPS 15: 170
23. Casey PJ, Gilman AG (1988) J Biol Chem 263: 2577
24. Ahmadian MR, Mittal R, Hall A, Wittinghofer A (1997) FEBS Lett 408: 315
25. Freissmuth M, Waldhoer M, Bofill-Cardona E, Nanoff C (1999) Trends Pharmacol Sci 20: 237
26. Takai Y, Takuya S, Takashi M (2001) Physiol Rev 81: 153
27. Berridge MJ (1993) Nature 361: 315
28. Berrige MJ, Irvine RF (1984) Nature 312: 315
29. Boarder MR (1994) TIPS 15: 57
30. Eichholtz T, Jalink K, Fahrenfort I, Molenaar WH (1993) Biochem J 291: 677
31. Brown AM (1991) News Phys Sci 6: 158
32. Strunecka A, Patocka J (2002) Aluminofluoride complexes: A useful tool in laboratory investigations, but a hidden danger for living organisms? In: Shapiro P, Atwood D (eds), Group 13 Chemistry: Fundamental research, material science and catalysis. ACS Vol., Washington
33. Bartus RT, Dean RL, Beer B, Lippa AS (1982) Science 217: 408
34. Bowen DM (1983) Banbury Rep V15: 219
35. Price DL, Whitehouse PJ, Struble RG, Price DL, Cork IC, Hedreen JC, Kitt CA (1983) Basal forebrain cholinergic neurons and neuritic plaques in primate brain. In: Katzman R (ed), Biological aspects of Alzheimer 's disease. Cold Spring Harbor Laboratory, Cold Spring Harbor, New York, p 65
36. Shimohama S, Taniguchi T, Fujiwara M, Kameyama M (1986) J Neurochem 46: 288
37. Vidal C, Changeux JP (1996) News Physiol Sci 11: 202
38. Mesulam MM (1986) News Physiol Sci 1: 107
39. Americ SP, William M (1995) Neuronal nicotinic acetylcholine receptors: novel targets for CNS therapeutics. In: Psychopharmacology: the fourth generation of progress. Raven Press, New York, p 1001

40. Qian Z, Drewes LR (1989) J Biol Chem 264: 21720
41. Ferrari-Di LG, Flynn DD (1993) Life Sci 53: 439
42. Wallace MA, Claro E (1993) Neurochem Res 18: 139
43. Inestrosa NC, Perelman A (1987) Trends Pharmacol Sci 10: 325
44. Platt B, Fiddler G, Riedel G, Henderson Z (2001) Brain Res Bull 55: 257
45. Ellman GL, Courtney KD, Andres V, Featherstone RM (1960) Biochem Pharmacol 7: 88
46. Zatta P, Zambenedetti P, Bruna V, Fillipi B (1994) NeuroReport 5: 1777
47. Soininen H, Lehtovirta M, Helisalmi S (1995) NeuroReport 6: 2518
48. Inestrosa NC, Alvarez A, Perez CA (1996) Neuron 16: 881
49. Rana RS, Hokin LE (1990) Physiol Rev 70: 115
50. Hokin LE, Hokin MR (1958) J Biol Chem 233: 818
51. Hokin MR (1970) J Neurochem 17: 357
52. Simmonds SH, Strange PG. (1985) Neurosci Lett 60: 267
53. Berridge MJ, Downess CP, Hanley MR (1982) Biochem J 206: 587
54. Chandler LJ, Crews FT (1990) J Neurochem 55: 1022
55. Candura SM, Castoldi AF, Manzo L, Costa LG (1992) Neurochem Res 17: 1133
56. Blackmore PF, Lynch CJ, Uhing RJ, Fitzgerald T, Bocckino SB, Exton JH (1988) Adv Exp Med Biol 232: 169
57. Nadakavukaren JJ, Welsh DK, Peppert SM (1990) Brain Res 507: 181
58. Shimohama S, Homma Y, Suenaga T, Fujimoto S, Taniguchi T, Araki W, Yamaoka Y, Takenawa T, Kimura J (1991) Am J Pathol 139: 737
59. Ripova D, Nemcova V, Höschl C, Fales E, Majer E, Strunecka A (1998) Lipid composition of different brain regions in patients with Alzheimer's disease and multi-infarct dementia. In: Fisher A, Yorshida M, Hanin I (eds), Progress in Alzheimer's and Parkinson's diseases. Plenum Press, New York, p 301
60. Khachaturian ZS, Cotman CW, Pettergrew JW (eds) (1989) Calcium, membranes, aging, and Alzheimer's disease. New York Academy of Sciences Vol 568, New York
61. Khachaturian ZS (1998) Am J Med 104: 26S
62. Thibault O, Porter NM, Chen KC, Blalock EM, Kaminker PG, Clodfelter GV, Brewer LD, Landfield PW (1998) Cell Calcium 24: 417
63. Verkhratsky A, Toescu EC (1998) Trends Neurosci 21: 2
64. Ripova D, Platilova V, Strunecka A, Jirak R, Höschl C. (2000) Neurobiol Aging 21: 729
65. Gibson GE, Zhang H, Toral-Barza L, Szolosi S, Tofel-Grehl B (1996) Biochim Biophys Acta 1316: 71
66. Arispe N, Pollard HB, Rojas E (1994) Mol Cell Biochem 140: 119
67. Gandolfi L, Stella MP, Zambenedetti P, Zatta P (1998) Biochim Biophys Acta 1406: 315
68. Strunecka A, Ibrahim N El-Desouki, Palecek J, Kmonickova E, Krpejsova L, Potter BV (1991) Receptor Humana Press 1: 141
69. Chen Y, Penington NJ (2000) J Neurophysiol 83: 1273
70. Mandelkow EM, Mandelkow E (1998) Trends Cell Biol 8: 425
71. Carlier MF, Didry D, Melki R, Chabre M, Pantaloni D (1988) Biochemistry 27: 3555
72. Côté M, Payet MD, Dufour MN (1997) Endocrinology 138: 3299
73. Caporaso GL, Gandy SE, Buxbaum JD (1993) Proc Natl Acad Sci USA 89: 3055
74. Lunardi J, Dupuis A, Garin J, Issartel JP, Michel L, Chabre M, Vignais PV(1988) Proc Natl Acad Sci USA 85: 8958
75. Missiaen L, Wuytack F, De Smedt H, Vrolix M, Casteels R (1988) Biochem J 253: 827
76. Selbie LA, Hill SJ (1998) Trends Pharmacol Sci 19: 87
77. Shafer T, Mundy WR, Tilson HU (1993) Brain Res 629: 133
78. Hattersley JG (1999) J Orthomol Med 14: 1
79. Xie ZM, Ye ZH, Wong MH (2001) Environ Int 26: 341
80. Flaten TP, Odegard M (1988) Food Chem Toxic 26: 959
81. Nabrzyski M, Gajewska R (1995) Z Lebensm Unters Forsch 201: 307
82. Martin RB (1988) Biochem Biophys Res Commun 155: 1194
83. Stauber JL, Florence TM, Davies CM, Adams MS, Buchanan SJ (1999) J AWWA 91: 84

84. Greger JL, Sutherland JE (1997) Crit Rev Lab Sci 34: 439
85. Alfrey AC, Heggh A, Craswel L (1980) Am J Clin Nutr 33: 1509
86. Fernandez-Lorenzo JR, Cocho JA, Rey-Goldar ML, Couce M, Fraga JM (1999) J Pediatr Gastroenterol Nutr 28: 270
87. Kissa E (1987) Clin Chem 33: 253
88. Bito H, Ikeda K (1994) Anesth Analg 79: 946
89. Matsumura C, Kemmotsu O, Kawano Y, Takita K, Sugimoto H, Mayumi T (1994) J Clin Anesth 6: 419
90. Ziegler H, Hauck K, Hefti AF (1991) Schweiz Monatsschr Zahnmed 101: 1529
91. Koparal E, Ertugrul F, Oztekin K (2000) J Clin Pediatr Dent 24: 299
92. Pak CYC, Sakhaee K, Gallagher C, Parcel C, Peterson R, Zerwekh JE, Lemke M, Britton F, Hsu MC, Adams B (1986) J Bone Miner Res 1: 563
93. Oliveby A, Lagerlof F, Ekstrand J, Dawes C (1989) Caries Res 23: 243
94. Kono K, Yoshida Y, Watanbe M, Orita Y, Dote T, Bessho Y (1993) Int Arch Occup Environ Health 65: 95
95. Ehrnebo M, Ekstrand J (1986) Int Arch Occup Environ Health 58: 179
96. Arnow PM, Bland LA, Houchins GS, Fridkin S, Fellner SK (1994) Annu Intern Med 121: 339
97. Ferry G (1989) New Scientist 27: 5
98. Lote CJ, Saunders H (1991) Clin Sci 81: 289
99. Walton J, Tuniz C, Fink D, Jacobsen G, Wilcox D (1995) Neurotoxicology 16: 187
100. Rubanyi GM, Polokoff MA (1994) Pharmacol Rev 46: 325
101. Pirotton S, Erneux C, Boeynaemes JM (1987) Biochem Biophys Res Commun 147: 1113
102. Magnusson MK, Halldorson H, Kjeld M, Thorgeirsson G (1989) Biochem J 264: 703
103. Gerritsen ME, Mannix RJ (1990) Adv Exp Med Biol 275: 115
104. Whatley RE, Fennell DF, Kurrus JA, Zimmerman GA, McIntyre TM, Prescott SM (1990) J Biol Chem 265: 15550
105. Boonen HC, De Mey JG (1990) Arch Pharmacol 342: 462
106. La Belle EF, Murray BM (1990) FEBS Lett 268: 91
107. Ratz PH, Blackmore PF(1990) J Pharmacol Exp Ther 254: 514
108. Varner JA, Jensen KF, Horvath W, Isaacson RL (1998) Brain Res 784: 284
109. Zhou J, Sims C, Chang CH, Mattera BL, Hopfer U, Douglas J (1990) Proc Natl Acad Sci USA 87: 7532
110. Kremer S, Harper P, Hagele R, Skorecki K (1988) Can J Physiol Pharmacol 66: 43
111. Brunskill NJ, Morrissey JJ, Klahr S (1992) Kidney Int 42: 11
112. Borke JL, Whitford GM (1999) J Nutr 129: 1209
113. Allgeier A, Offermanns S, Van Sande J, Spicher K, Schultz G, Dumont JE (1994) J Biol Chem 269: 13733
114. Bubis M, Zisapel N (1995) Neurochem Int 27: 177
115. Barrett P, Morris M, Choi WS, Ross A, Morgan PJ (1999) Biol Signal Recept 8: 6
116. Morgan PJ, Hastings MH, Thompson M, Barrett P, Lawson W, Davidson G.J (1991) Mol Endocrinol 7: 137
117. Nadakavukaren JJ, Welsh DK, Peppert SM (1990) Brain Research 507: 181
118. Luke J (2001) Caries Res 35: 125
119. Suarez-Fernandez MB, Soldado AB, Sanz-Medel A, Vega JA, Novelli A, Fernandez-Sanchez MT (1999) Brain Res 835: 125
120. Loweth AC, Williams GT, Scarpello JH, Morgan NG (1996) Exp Cell Res 229: 69
121. Conroy LA, Jenkinson EJ, Owen JJ, Michell RH (1995) Eur J Immunol 25: 1828
122. Spittle B (1994) Int Clin Psychopharm J 9: 79
123. Waldbott GL, Burgstahler AW, McKinney HL (1978) Fluoridation: the great dilemma. Coronado Press, Lawrence, Kansas
124. Li XS, Zhi JL, Gao RL(1995) Fluoride 28: 189
125. Spittle B (2000) Fluoride 33: 49
126. Lu Y, Sun ZR, Wu LN, Wang X, Lu W, Liu SS (2000) Fluoride 33: 74
127. Roush W (1996) Science 271: 1056

# Fluoroaluminate Chemistry

B. Conley[1], D.A. Atwood[2]

Department of Chemistry, University of Kentucky, Lexington, KY 40506, USA
[1] *E-mail: phoibos.mousagetes@prodigy.net*
[2] *E-mail: datwood@pop.uky.edu*

Salts of the tetrafluoroaluminate (TFA) anion have been known for many years. Yet aqueous routes to pure materials have not been established despite the substantial amount of literature available on the subject. The present review will discuss the various synthetic methods that have been employed and will demonstrate a revised method for preparing pure TFA salts. These materials are important because TFA is able to stimulate various guanosine nucleotide-binding proteins (G-proteins) and inhibit P-type ATPases by serving as a non-hydrolyzing phosphate mimic. Additionally, various TFA salts serve as precursors to aluminum trifluoride, which is used as a catalyst for chlorofluorocarbon isomerizations and fluorinations.

**Keywords:** Tetrafluoroaluminate, Fluoroaluminates, Aluminates, G-proteins, Phosphate analogue



Structure and Bonding, Vol. 104

## List of Abbreviations

| | |
|---|---|
| TFA | tetrafluoroaluminate |
| GDP | guanosine diphosphate |
| GTP | guanosine triphosphate |
| cAMP | cyclic adenosine monophosphate |
| NDP | nucleoside 5′-diphosphate |
| ATFA | ammonium tetrafluoroaluminate |
| MATFA | tetramethylammonium tetrafluoroaluminate |
| B15C5 | benzo-15-crown-5 |
| PS | 1,8-bis(*N,N*-dimethylamino)naphthalene |
| coll | *sym*-collidine |

## 1 Introduction

Aluminum is the most abundant metal in the earth's crust while fluorine is ranked thirteenth among the elements [1]. As a hard metal ion, aqueous $Al^{3+}$ interacts most strongly with hard donors, and therefore complexation with $O^{2-}$ and $F^-$ donor ligands produces very stable compounds. Aluminum forms stronger complexes with fluoride than with the other halides and the Al—F bond strength is approximately 655 kJ $mol^{-1}$. Out of 60 metal species, $Al^{3+}$ is surpassed only by $Sc^{3+}$ in forming the strongest bond to fluorine [2]. The effect of the bond strength can be observed in the extremely high lattice energy of $AlF_3$, which is the most thermodynamically stable compound of aluminum and fluorine. This accounts for some of the difficulty in obtaining tetrafluoroaluminates in a pure form. Accordingly, caution must be exercised when mixing reagents containing Al and F. Yet surprisingly, there are few minerals containing these two minerals. Cryolite, $Na_3AlF_6$, is of limited natural distribution and is found mostly in one location on the west coast of Greenland [3]. Likewise, rare minerals resulting from cryolite weathering such as Thomsenolite and Pachnolite (both of formula $NaCaAlF_6 \cdot H_2O$) can be found [4]. Muscovite, $KAl(SiO_4)(OH)F$, is more widespread but still is not sufficiently abundant to be commercially useful except as an ornamental mica [5].

The commercially utilized Al-F compounds are inorganic and synthetic in origin. Thus, cryolite ($Na_3AlF_6$), used on a substantial scale (millions of tons per year), can be synthesized from soda, alumina, and hydrofluoric acid [6]. Preparations of tetrafluoroaluminate salts are executed mostly in situ due to the small amounts of material needed for the various applications. $AlF_3$ is a catalyst for various reactions (see below) and can be prepared, by the thermolysis of $[AlF_4^-]$-containing salts [7]. More detailed and specific syntheses were subsequently developed, as described below.

Aluminum trifluoride is important in the industrial production of aluminum metal, as it increases the conductivity of electrolytes in the electrolytic process [6]. Aluminum trifluoride is also used as a catalyst for chlorofluorocarbon isomerization and fluorination. High surface area $AlF_3$ dispersed onto carbon, organic, or inorganic supports may be a useful catalyst for these or other related reactions [8]. For instance, a fluoroaluminum species generated on a support of alumina was treated with chromium to obtain active olefin polymerization catalysts [9].

When soldering aluminum, potassium tetrafluoroaluminate can be used to remove oxides and other disruptive covering films on the metal surface [10]. A mix of fluoroaluminates (including $KAlF_4$ as well as $KAlF_6$) allows brazing of aluminum at temperatures lower than what could be accomplished with fluoride fluxes [11]. Potassium-cesium tetrafluoroaluminate has similar applications [12]. $LiAlF_4$ can be incorporated into the carbon cathode current collector of non-aqueous lithium batteries. It is also used as a non-aqueous electrolyte that can be dispersed throughout the cathode collector [13].

## 2 Biological Activity

Aluminum fluoride complexes, especially tetrafluoroaluminates (TFA), are currently receiving intense scrutiny due to their potential biological activity [14]. In particular, the ability of TFA (or $AlF_3 \cdot H_2O$) to act as phosphate analogues and thereby stimulate various guanosine nucleotide-binding proteins (G-proteins) [15] and inhibit P-type ATPases [16]. G-proteins fulfill an integral role in an enormous variety of biological signaling mechanisms and thereby help to control myriad important life processes. As a consequence of the ubiquity of G-proteins, tetrafluoroaluminates are now widely used in laboratory studies to investigate the physiological and biochemical changes effected in cellular systems by aluminofluoride complexes.

For most of the biochemical and physiological studies involving the putative $AlF_4^-$ anion, the fluoride source is usually sodium fluoride and the aluminum source is aluminum nitrate or aluminum chloride. Solutions are usually prepared with millimolar concentrations of sodium fluoride and micromolar concentrations of the aluminum source.

Transfer of phosphate groups is the basic mechanism in regulation of the activity of numerous enzymes involved in energy metabolism, cell signaling, cell motility, and regulation of cellular growth. TFA anion acts as a high-affinity analogue of the $\gamma$-phosphate of phosphonucleosides (Fig. 1) [17, 18]. It can mimic the role of the $\gamma$-phosphate only if the $\beta$-phosphate is present and remains unsubstituted. The effect is more readily seen with G-proteins because

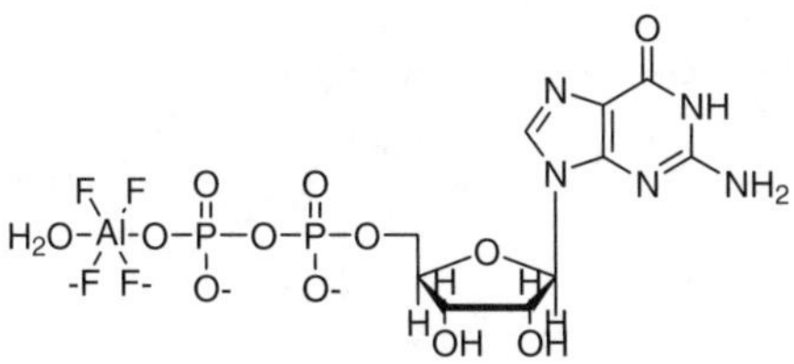


**Fig. 1.** Model with TFA as $\gamma$-phosphate analogue (one F atom may be $H_2O$)

guanosine diphosphate (GDP) is always tightly bound at the active site after the hydrolysis of guanosine triphosphate (GTP). An X-ray crystal structure of the bound tetrafluoroaluminate complex with GDP obtained from the $\alpha$ subunit of the heterotrimeric G-protein transducin shows the importance of water molecules to the integrity of the bound complex, and is illustrative not only of the association of the TFA anion with the GDP $\beta$-phosphate, but also of the extensive hydrogen bonding between the fluorine atoms and the neighboring amino acid residues of the G-protein (Fig. 2) [19]. These interactions contribute greatly to the stability of the complex and account for the greater binding affinity of the anion compared to that of orthophosphate.

The tetrafluoroaluminate complex was proposed to act as an analogue of the terminal phosphate of GTP because the Al—F bond length is close to the phosphate P—O bond length, and because the TFA and $PO_4^{3-}$ structures are both tetrahedral. Fluorine and oxygen have nearly the same size and the same valence orbitals, and aluminum and phosphorus have their valence electrons in the same third shells. The two bonding schemes differ, however, in that the former is more ionic while the latter is more covalent. That is, in phosphate, oxygen is bound covalently to the phosphorus and does not exchange with oxygen from solvent, while in $AlF_4^-$ the bonding between the electropositive aluminum and the highly electronegative fluorine is more ionic in character, allowing fluorine in the bound complex to exchange freely with fluoride ions in solution.

Fluoroaluminate complexes can mimic the action of many hormones, neurotransmitters, and growth factors. G-protein-mediated cell responses are key steps in neurotransmission and intercellular signaling in the brain [20], and TFA acts as an active stimulatory species [21]. Exposure of osteoblasts to TFA results in a marked potentiation of intracellular orthophosphate transport, alluding to the anion's ability to increase bone mineralization [22]. Brief exposure to aluminum fluoride complexes induces prolonged enhancement of synaptic transmission [23] and can potentially affect the activity of many other ion channels and enzymes in the kidney [24]. Rapid and dynamic changes of the cytoskeletal actin network are of vital importance to the motility of many cells, and TFA induction effects a pronounced and sustained

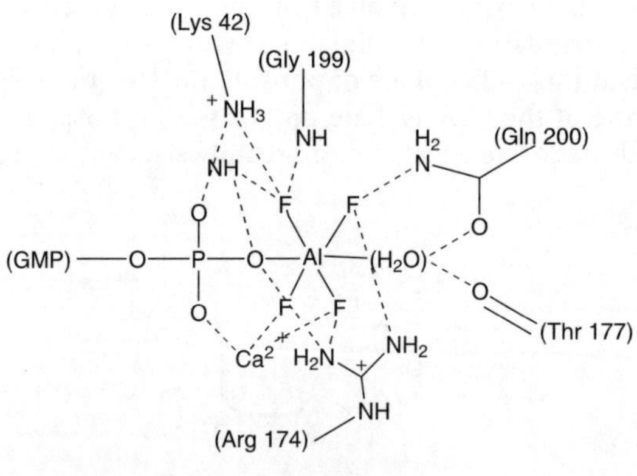


**Fig. 2.** H-bonding in bound GDP-TFA in transducin

increase in a filamentous form of actin in intact human neutrophils [25]. TFA has also been shown to mimic the activity of hormones that mobilize $Ca^{2+}$ into cytosol from intracellular stores, effectively decreasing cellular levels of the second messenger molecule cyclic adenosine monophosphate (cAMP) [26, 27].

It should be noted that the human body does possess natural barrier systems to prevent aluminum intake. There are various physiological ligands, such as transferrin, citrate, and silicilic acid, which are efficient buffers in preventing the intake of aluminum under natural conditions [28]. Yet the formation of TFA requires only trace amounts of aluminum, and the bioavailability of aluminum to living organisms is increasing. Fluorine, of course, is readily available in drinking water.

There is some uncertainty over the identity of the biologically active aluminum species. Species with four fluorines, six fluorines, and with fluorines replaced by hydroxides all appear possible. The distribution of these mixed fluoroaluminate and fluorohydroxyaluminate species seems highly dependent upon prevailing solution pH and pF [29]. For example, multinuclear NMR spectroscopy was used to study the ternary system $Al^{3+}$, $F^{-}$, and NDP (nucleoside-5′-diphosphate) in aqueous solution at pH 6 without protein [30, 31]. Ternary complexes (NDP)$AlF_x$ ($x=1$–3) were found, but no (NDP)$AlF_4$ was detected. Further multinuclear NMR studies of fluoroaluminate species in aqueous solutions over a wider pH range (2–8) with varying $[F][Al]^{-1}$ concluded that all the fluoroaluminate complexes observed in aqueous solution are hexacoordinated with an octahedral geometry [32]. Further research aimed specifically at elucidating the nature of the aluminum phosphate interaction will certainly provide the answer in the near future.

## 3 Aqueous Syntheses

### 3.1 Sodium Tetrafluoroaluminate (NaTFA)

Historically, many methods have been used to synthesize the tetrafluoroaluminate (TFA) anion with varying degrees of purity [33]. Indeed, aluminofluoride complexes are formed spontaneously in water containing fluoride and trace amounts of aluminum. These methods, however, produce mixtures of aluminates. This partially explains why most applications employing this anion have utilized in situ syntheses. Not surprisingly, there was some controversy over the actual products obtained from early syntheses of TFA salts [34]. The first of these methods involved the partial neutralization of $H_3AlF_6$ at 70–80 °C [Eq. (1)] [35]. The product was found to be soluble in water. A second proposed method involved the combination of $AlF_3$ and NaF at 80 °C [Eq. (2)] [36]. This reaction did not produce TFA but two other products, $AlF_3 \cdot 3\,H_2O$ and $AlF_3 \cdot 1.5\,NaF$ (chiolite), instead. In an effort to explain this earlier work and to determine the water solubility of NaTFA, a more stringently controlled synthesis was devised [37]. Thus, HTFA was prepared *in situ* at 80 °C (higher temperatures drive the equilibrium toward more HTFA) [38] and then neutralized to a residual acidity of ~0.2% HF [Eq. (3)]. The resulting NaTFA precipitate was isolated by filtration, washed with alcohol, and dried at 105 °C. It was subsequently determined that NaTFA decomposes to chiolite and other species when dissolved in water at 25 °C. At 75 °C, however, the equilibrium is shifted toward maintenance of NaTFA, and decomposition does not occur. The solubility in water at this temperature is 0.133% [38].

A seemingly straightforward route to aqueous NaTFA is the combination of an aluminum salt and a fluoride source [Eq. (4)]. As alluded to previously, this method, utilizing millimolar concentrations of the fluoride source and micromolar concentrations of the aluminum salt, is the one most often employed for biological studies [14]. Yet the solution thus obtained actually contains an equilibrium mixture of the various possible fluorinated species [39].

$$H_3[AlF_6](aq) + NaOH(aq) \rightarrow Na[AlF_4] \cdot H_2O \text{ (putative)} \tag{1}$$

$$NaF(aq) + AlF_3(aq) \rightarrow AlF_3(H_2O)_3 + NaF \cdot AlF_3 \tag{2}$$

$$HF + Al(OH)_3 + Na_2CO_3 \rightarrow Na[AlF_4](s) \tag{3}$$

$$4\,NaF + Al(NO_3)_3 \rightarrow Na[AlF_4] + 3\,NaNO_3 \tag{4}$$

## 3.2 Potassium Tetrafluoroaluminate (KTFA)

In contrast to the preparations for NaTFA, the preparation of KTFA appears much more certain. KTFA can be prepared from a combination of aluminum hydroxide, hydrofluoric acid, and potassium hydroxide [Eq. (5)]. The salt could be isolated as a precipitate and dried under vacuum at 80 °C. It was found to have a melting point of 546–550 °C by differential scanning calorimetry [40].

$$Al(OH)_3 + 5\,HF + 2\,KOH \rightarrow K[AlF_4] + 5\,H_2O + KF \tag{5}$$

## 3.3 Ammonium Tetrafluoroaluminate (ATFA)

The preparation of pure ATFA has been attempted in a number of ways. A "wet" method involves adding an inorganic acid such as $H_2SO_4$ to an aqueous slurry of $(NH_4)_3AlF_6$ and either $Al(OH)_3$ or $Al_2O_3$, leading to the precipitation of ATFA [Eq. (6)] [41]. The reaction system was maintained under atmospheric pressure and at 70–100 °C. To neutralize the free ammonia in the slurry and maintain the solubility of the ATFA, the inorganic acid was added in such an amount that the pH of the slurry after completion of the reaction was 4 to 7. The particle size of the ATFA can be controlled by varying the rate at which the inorganic acid is added to the slurry. For example, when the total volume of the acid is poured into the slurry at one time, the particle size of the crystalline ATFA ranges from approximately 10–20 μm. Particle size increases to a range of 20–50 μm when the acid is added intermittently. It should be noted that no characterization data were provided to support this preparation. In attempts to replicate the procedure it was found that ATFA was indeed produced, but as a mixture with other aluminum-containing species (as indicated by solid-state $^{27}$Al-NMR) [39].

ATFA of sufficient purity to be used as a precursor for the formation of $\gamma$-$AlF_3$ can be obtained from the coprecipitation of aqueous solutions of aluminum nitrate and ammonium fluoride followed by aging at 100 °C [Eq. (7)] [42]. The precipitate was then filtered, washed with ethanol, and dried at 100 °C for 24 hours. Various permutations of this synthesis could also be employed.

A simple preparation of ATFA has been employed recently using ammonium bifluoride and aluminum nitrate [Eq. (8)] [43]. Dissolution of each reactant in water followed by mixing of the solutions results in formation of a white precipitate. The precipitate is then filtered and triturated with acetone. There is no need to heat the precipitate, as it dries completely in a short amount of time at room temperature [39].

$$4\,(NH_4)_3AlF_6 + 2\,Al(OH)_3 + 3\,H_2SO_4 \rightarrow 6\,NH_4AlF_4 + 3\,(NH_4)_2SO_4 + 6\,H_2O \quad (6)$$

$$Al(NO_3)_3 \cdot 9\,H_2O + 4\,NH_4F\ (\text{dropwise}) \rightarrow NH_4AlF_4 + 3\,NH_4NO_3 \quad (7)$$

$$2\,NH_4F \cdot HF + Al(NO_3)_3 \cdot 9\,H_2O \rightarrow [NH_4][AlF_4] + NH_4NO_3 + 2\,HNO_3 \quad (8)$$

### 3.4 Tetramethylammonium Tetrafluoroaluminate (MATFA)

MATFA was originally prepared as the product obtained from the dissolution of aluminum trifluoride in aqueous HF in the presence of tetramethylammonium fluoride [44]. Dehydration of the resulting precipitate below 120 °C produces hygroscopic MATFA.

## 4 Anhydrous Syntheses

Perhaps the most simple anhydrous method for preparing ATFA is the heating of $(NH_4)_3AlF_6$ at 175 °C [Eq. (9)] [45]. Pure ATFA was claimed based upon XRD (X-ray powder diffraction) analysis, which is only capable of detecting crystalline phases. A preparation for ATFA claiming a 99% yield based upon XRD analysis involves the use of ammonium bifluoride in combination with an amine complex of aluminum trichloride (presumably to prevent hydrolysis of aluminum) in toluene [Eq. (10)] [46]. After completion of the reaction the toluene was distilled off and the product was purified by washing with water to remove $NH_4Cl$, and was subsequently dried in an oven.

TFA with organic base countercations other than ammonium or tetramethylammonium can be prepared by evaporating a mixture of $Al(OH)_3$ dissolved in HF and the appropriate organic base [47]. Organic bases such as hydroxylamine, pyridine, quinoline, morpholine, 2-aminopyridine, and $\alpha,\alpha'$-bipyridyl could be employed as ammonium counterions to TFA. The corresponding fluoroaluminates were highly hygroscopic solids. Thermal decomposition of these compounds, in general, produced a mixture of aluminum fluoride and oxide, indicating the presence of either hydrated [BaseH][TFA] or other oxygen-containing compounds.

TFA salts can be extracted from mixtures of aluminum and sodium fluoride into dimethyl sulfoxide (DMSO) and acetonitrile solutions containing benzo-15-crown-6 (B15C5) [48]. This solubilizes, and apparently stabilizes, salts having the general formula [M(B15C5)][TFA] [$M = Na^+$ or 1,8-bis(dimethylamino)naphthalene]. Varying the concentration of B15C5 from 0.2 to 0.7 M both in DMSO and in acetonitrile, as well as increasing the NaF:Al ratio from 4:1 to 50:1, had no significant effect on the extraction. In DMSO, however, six-coordinate aluminum fluoride complexes were also present.

Cation exchange has not proven to be a useful means by which to vary the countercation of TFA. For instance, it is not possible to replace a univalent cation with a divalent cation (for two TFA units) since the divalent anion, $AlF_5^{2-}$, is formed readily. This is demonstrated in the synthesis of [Mg] $[AlF_5^{2-}] \cdot 2.2\ H_2O$ [49].

The addition of trimethylaluminum to pyridinium fluoride produced the first anhydrous TFA product [Eq. (11)] [50]. This compound could subsequently be used to prepare an extensive range of other anhydrous salts through cation exchange [8]. Thus, the compound [PS][TFA] [PS = Proton Sponge®, 1,8-bis(dimethylamino)naphthalene] was prepared in a glove box by slurrying [pyridineH][TFA] into a solution containing excess PS dissolved in dry acetonitrile. Likewise, slurrying the pyridinium derivative in neat *sym*-collidine (2,4,6-trimethylpyridine, henceforth "coll") in a glove box and heating at 120 °C for 30 minutes produced [collH][TFA].

The compound tetraphenylphosphonium TFA ($[Ph_4P][TFA]$) was prepared by cation exchange with the [collH] salt dissolved in methanol and $Ph_4PBr$. The product was recrystallized from hot acetone or acetonitrile. The derivative $[Ph_4As][TFA]$ was prepared using $Ph_4AsCl$ and was characterized in a similar manner. Another synthesis involved mixing the [collH] salt with $Me_4NCl$ in dry methanol [51]. The by-product, [collH][Cl], was removed by sublimation in flowing nitrogen below 200 °C, and the salt $[Me_4N][TFA]$ was left behind. An alternate route to these types of compounds involves the addition of a bifluoride to an alkylaluminum reagent in toluene and heating at 80 °C for 2 hours [Eq. (12)] [52]. The product is then isolated by crystallization upon cooling.

The $\beta$-phase of ATFA could be obtained by heating [pyridineH][TFA] under $N_2$ to approximately 180 °C in formamide [Eq. (13)] [53]. Pyridine is evolved from the solution and HTFA is presumed to remain behind. It was subsequently found that the compound was actually $[pyridineH][Al_3F_{10}]$

[54]. This combines with the formamide solvent to eliminate carbon monoxide.

$$(NH_4)_3AlF_6 \rightarrow NH_4AlF_4 + 2\,NH_4F \tag{9}$$

$$AlCl_3 \cdot N(Me)_3 + 2\,NH_4HF_2 \rightarrow NH_4AlF_4 + (Me)_3N \cdot HCl + HCl \tag{10}$$

$$Me_3Al + 4\,HF \cdot \text{pyridine} \rightarrow 4\,CH_4 + [\text{pyridineH}][AlF_4] \tag{11}$$

$$[Ph_4P] \cdot HF_2 + [Ph_4P][(iBu)_2AlF_2] \rightarrow [Ph_4P][AlF_4] \tag{12}$$

$$C_6H_5NH\ AlF_4 + NH_2CHO \rightarrow NH_4AlF_4 + C_6H_5N + CO \tag{13}$$

# 5 Characterization

## 5.1 Spectroscopic

The IR spectra for these compounds showed a sharp band at approximately 780 $cm^{-1}$ attributed to the Al-F stretching frequencies. For example, this band in the compound $[Ph_4P][TFA]$, with discrete, unassociated TFA units, appeared at 783 $cm^{-1}$ [52] and for [PSH][TFA] it appeared at 785 $cm^{-1}$ [50]. The Raman spectra, when reported, appeared as a sharp band at approximately 630 $cm^{-1}$. Mixtures presumed to contain some TFA show IR bands in the range of 410–675 $cm^{-1}$ [47, 51]. IR bands may be observed from 567–825 $cm^{-1}$ for hydrated TFA anions with inorganic cations ($Na^+$, $Li^+$, $K^+$, $Rb^+$, $Cs^+$, and $Tl^+$) [39].

NMR peaks in the range of –187 to –194 ppm ($^{19}$F-NMR) and 49 to 52 ppm ($^{27}$Al-NMR) were observed for the TFA salts prepared via anhydrous routes. Thus, they correspond to tetrahedral aluminum coordinated by four fluorine atoms. This is demonstrated, for example, in the coupling for [PSH][TFA] in $CD_3CN$, which was observed as a sextet from $^{19}$F-$^{27}$Al in the $^{19}$F-NMR, and a quintet from $^{27}$Al-$^{19}$F in the solid state $^{27}$Al-NMR.

In aqueous solutions, however, there is no evidence for the existence of discrete four-coordinate TFA units. Rather, a rapid pH-dependent exchange occurs between $H_2O$, $OH^-$, and $F^-$ ligands binding to the aluminum cation based upon the $^{27}$Al- and $^{19}$F-NMR data [55]. Specific findings include the existence of complexes with one or two $OH^-$ groups per aluminum with up to three fluorines undergoing rapid exchange. All solution species were six-coordinate and octahedral. Furthermore, after the replacement of a coordinated water by fluoride {on the cation, $[Al(H_2O)_6]^{3+}$, for example} the exchange rate of the remaining coordinated waters increases by two orders of magnitude [56]. These observations are in keeping with the theoretical results, excluding the possibility of tetrahedral TFA in aqueous solution [57].

## 5.2 Structural

Aluminum halide chemistry presents several examples of tetrahedral $[AlX_4]$ species where X = Cl, Br, or I [58–60]. Yet the existence of discrete, tetrahedral species for fluorine had for many decades been controversial due to the lack of structural proof. This is despite the indirect evidence for TFA (as presented previously) as well as studies of molten salts [61–65].

In the structure of [PSH][TFA] (Fig. 3) [50] the expected tetrahedral anion was confirmed. Not surprisingly, the F-Al-F bond angles were approximately 109°, with equivalent Al-F distances of around 1.62 Å. The closest contact between the coordinated fluoride and the chelated proton of the cation was 2.77 Å, which could be considered a long hydrogen bond. The crystal structures of $[Ph_4P]$[TFA], $[Ph_4As]$[TFA], and $[n\text{-}Bu_4N]$[TFA] (Fig. 4) have also been reported and are similar to that of [PSH][TFA] with discrete cations and anions [8].

The structure of [collH][TFA] (Fig. 5) contains crystallographically independent TFA units and $[AlF_4]_\infty$ chains (also found in the structure of [pyridineH]$[Al_3F_{10}]$ [54]) with six-coordinate aluminum [50]. The $collH^+$ cations form strong hydrogen bonds to the terminal fluoride ions of the chain, effectively forming a sheath around these chains. Residing in a hydrophobic region between the chains defined by the $collH^+$ ions are two independent, discrete, tetrahedral $AlF_4^-$ anions. As before, the anions have no contact, other than Van der Waals interactions, with other species in the lattice. A monohydrate, $[Me_4N]_2[Al_2F_8(H_2O)_2]$, has also been structurally characterized

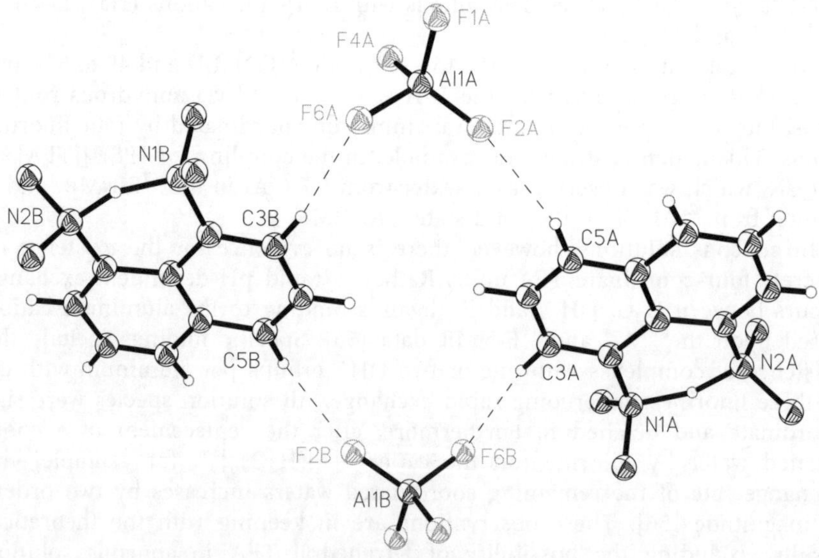


**Fig. 3.** Molecular structure of [PSH][TFA]

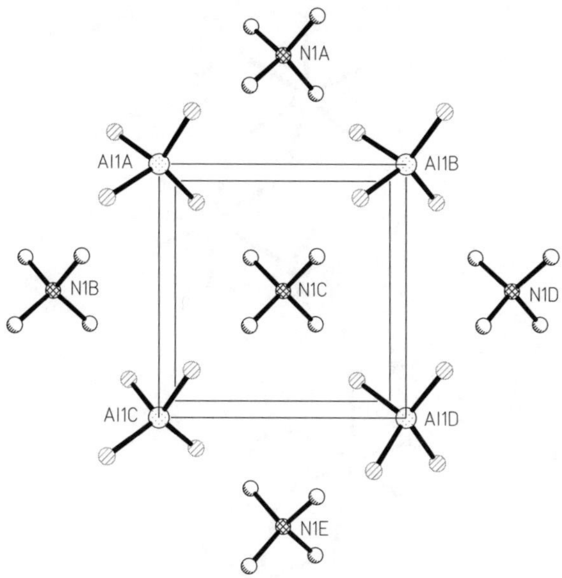


**Fig. 4.** Molecular structure of [*n*-$Bu_4N$][TFA] (only the α-carbons of the *n*-Bu groups are shown)

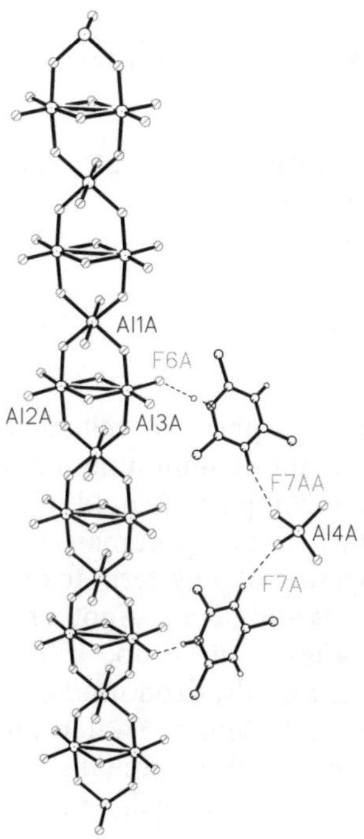


**Fig. 5.** Molecular structure of [collH][TFA] accentuating the individual $AlF_4$ units and the $AlF_6$ chains

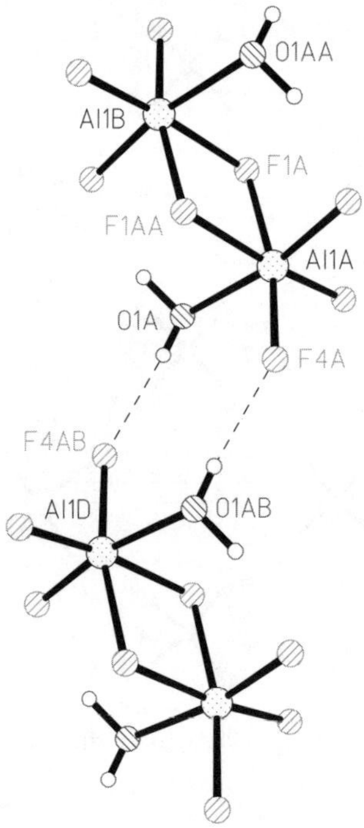


**Fig. 6.** Molecular structure of the anions of $[Me_4N][Al_2F_8(H_2O)_2]$

(Fig. 6). In the structure of a related compound, $AlF_3$ $(NH_3)_2$ (bridging F), only octahedral aluminum is observed [66].

## 6
## Conclusions

In view of the ubiquity of phosphate in cell metabolism together with the dramatic increase in the amount of aluminum and fluoride now found in our ecosystem, aluminofluoride complexes could represent a strong potential danger for living organisms, including humans [14]. One area of important research will be the investigation of long-term toxicological effects of exposure to TFA complexes on animals and plants. Another area of future research will be the determination of the relationship, if any, between aluminum in everyday products (cooking utensils, deodorants, antacids, food and beverage packaging), the increasing use of fluoride (water fluoridation, dental products, industrial fertilizers), and the health of humans.

Industrially, continued TFA research will be essential to the development of better $AlF_4^-$ precursors, which provide for cleaner decomposition to $AlF_3$ at lower temperatures. The deposition of $AlF_4^-$ onto organic and inorganic supports for subsequent decomposition to $AlF_3$ continues to be investigated.

Future research will also be conducted with the objective of preparing larger quantities of [M][TFA] (M = metal or organic cation) of greater purity with less dangerous starting materials, lower temperatures, and better cost-effectiveness. Preparation of higher quality fluxes containing TFA for the soldering of aluminum and aluminum alloys will also be an area of continual research and development.

All of these applications will benefit from fundamental synthetic and structural results such as those reported above. Thus, there will be a clear connection between laboratory research and commercial applications.

## 7 References


1. Greenwood NN, Earnshaw A (1998) Chemistry of the elements. 2$^{nd}$ edn, Butterworth-Heinemann, Woburn
2. Bond AM, Hefter GT (1980) Critical survey of stability constants and related thermodynamic data of fluoride complexes in aqueous solution. IUPAC Chemical Data Series 27. Pergamon, New York
3. Edwards JD (1930) In: Edwards JD, Frary FC, Jeffries Z (eds), Aluminum and its production. McGraw-Hill Book Company, Inc. New York, Vol. 1, chap VIII
4. Adhikesavalu D, Cameron TS, Knop O (1985) Can J Chem 63: 3322
5. Arostegui J, Irabien MJ, Nieto F, Sanguesa J, Zuluaga MC (2001) Clay Clay Miner 49: 529
6. Altenpohl DG (1988) Aluminum: technology, applications and environment. 6th edn, The Aluminum Association, Washington DC
7. LeBail A, Fourquet JL, Bentrup U (1992) J Solid State Chem 100: 151
8. Harlow RL, Herron N, Thorn DL (1999) US Patent 5,986,023
9. McDaniel MP, Klendworth DD, Johnson MM (1992) US Patent 5,171,798
10. Conn PJ, Bowling Jr. JH (2001) US Patent 6,244,497
11. Ono M, Hattori M, Itaya E, Yanagawa Y (2000) US Patent 6,010,578
12. Belt HJ, Sander R, Rudolph W (1999) US Patent 5,985,233 and references cited therein
13. Zajack Jr. WV, Bis FF, DeBold FC, Kowalchik LA, Barnes JA (1998) US Patent 5,714,279
14. Strunecka A, Patocka J (1999) Fluoride 32: 230
15. Sternweis PC, Gilman AG (1982) Proc Natl Acad Sci USA 78: 4888; Gilman AG (1987) Annu Rev Biochem 56: 615
16. Troullier A, Girardet JL, Dupont Y (1992) J Biol Chem 267: 22821 (and references cited therein)
17. Bigay J, Deterre P, Pfister C, Chabre M (1987) EMBO J 6: 2907
18. Chabre M (1990) TIBS 15: 6
19. Sondek J, Lambright DG, Noel JP, Hamm HE, Sigler PB (1994) Nature 372: 276
20. Rana RS, Hokin LE (1990) Physiol Rev 70: 115
21. Candura SM, Castoldi AF, Manzo L, Costa LG (1991) Life Sci 49: 1245
22. Moonga BS, Pazianas M, Alam AS, Shankar VS, Huang CL, Zaidi M (1993) Biochem Biophys Res Commun 190, 496
23. Publicover SJ (1991) Exp Brain Res 84: 680
24. Zhou J, Sims C, Chang CH, Mattera BL, Hopfer U, Douglas J (1990) Proc Natl Acad Sci USA 87: 7532
25. Bengtsson T, Sarndahl E, Stendahl O, Andersson T (1990) Proc Natl Acad Sci USA 87: 2921
26. Blackmore PF, Bocckino SB, Waynick LE, Exton JH (1985) J Biol Chem 260: 14477
27. Blackmore PF, Exton JH (1986) J Biol Chem 261: 11056
28. Wilhelm M, Jager DE, Ohnesorge FK (1990) Pharmacol Toxicol 66: 4

29. Antonny B, Chabre M (1992) J Biol Chem 267: 6710
30. Nelson DJ, Martin RB (1991) J Inorg Biochem 43: 37
31. Wang X, Simpson JH, Nelson DJ (1995) J Inorg Biochem 58: 29
32. Martinez EJ, Girardet JL, Morat C (1996) Inorg Chem 35: 706
33. See Dutta U, Conley BD, Gilliam AC, Yearwood BC, Selegue JP, Atwood DA (2001) In: ACS Symposium Series, ACS for a brief overview of the syntheses
34. See Korobitsyn AS, Permyakova TA, Kondakov VP, Ust'yantseva TA, Leont'eva IA, Shishkina ZI, Kalitina LN (1983) Zh Prikl Khim 56: 887 for a good discussion of the many attempts to make TFA
35. Yatlov VS (1937) Zh Obshch Khim 7: 2439
36. Kozlov YuA, Belova NV, Leont'eva IA, Bogachov GN (1965) In: Pozin ME (ed), Studies in the chemistry and technology of mineral salts and oxides. Ed Izd Nauka, Leningrad
37. Masalovich VM, Katorina OV, Korobitsyn AS, Permyakova TA, Kulikova SP (1990) Russ J Inorg Chem 35: 968
38. Masalovich VM, Korobitsyn AS, Permyakova TA (1988) Russ J Inorg Chem 33: 264
39. This was determined by attempts to reproduce the literature preparations: Conley B, Shaikh T, Atwood D (patent submitted 2002)
40. Belt HJ, Sander R, Rudolph W (1999) US Patent 5,985,233. A similar preparation is reported in Willenberg H, Hellberg KH, Zschiesche H (1984) US Patent 4,428,920
41. Aramaki M, Etsuo U (1977) US Patent 4,034,068
42. Alonso C, Morato A, Medina F, Guirado F, Cesteros Y, Salagre P, Sueiras JE, Terrado R, Giralt A (2000) Chem Mater 12: 1148
43. Bukovec P, Siftar J (1975) Monatsh Chem 106: 483
44. Ferbinteanu M, Roesky HW, Cimpoesu F, Atanasov M, Kopke S, Herbst-Irmer R (2001) Inorg Chem 40: 4947
45. Shinn D, Crocket D, Haendler H (1966) Inorg Chem 5: 1927. See also Kidde GE (1972) US Patent 3,694,150
46. Marlett E (1991) US Patent 5,045,300
47. Sengupta AK, Sen K (1979) Indian J Chem 17A: 107
48. Petrosyants SP, Maliarik MA, Tolkacheva EO, Tsivadze Ayu (1998) Main Group Chem 2: 183
49. Riley KW, Horne A (1986) Anal Chim Acta 182: 257
50. Herron N, Thorn DL, Harlow RL, Davidson F (1993) J Am Chem Soc 115: 3028
51. Herron N, Harlow RL, Thorn DL (1993) Inorg Chem 32: 2985
52. Bartsch R, Schmutzler R, Spiegel GU, Stelzer O (1987) J Fluorine Chem 36: 107
53. Harlow RL, Herron N (1995) US Patent 5,417,954
54. Harlow RL, Herron N, Li Z, Vogt T, Solovyov L, Kirik S (1999) Chem Mater 11: 2562
55. Martinez EJ, Girardet JL, Morat C (1996) Inorg Chem 35: 706
56. Yu P, Phillips BL, Casey WH (2001) Inorg Chem 40: 4750
57. Martin RB (1988) Biochem Biophys Res Commun 155: 1194
58. Zaworotko MJ, Cameron TS, Linden A, Sturge KC (1989) Acta Crystallogr C45: 996
59. Jones DEH (1972) J Chem Soc Dalton Trans 567
60. Kidd RG, Truax BR (1968) J Am Chem Soc 90: 6867
61. Gilbert B, Mamantov G (1975) J Chem Phys 62: 950
62. Gilbert B, Mamantov G, Begun GM (1974) Inorg Nucl Chem Lett 10: 1123
63. Gilbert B, Williams SD, Mamantov G (1988) Inorg Chem 27: 2359
64. Matwiyoff NA, Wageman WE (1970) Inorg Chem 9: 1031
65. Colton R, Eller PG (1989) Aust J Chem 42: 1605
66. Ketchum DR, Schimek GL, Pennington WT, Kolis JW (1999) Inorg Chim Acta 294: 200

# Author Index Volumes 101–104

# Subject Index

Printing (Computer to Film): Saladruck Berlin
Binding: Stürtz AG, Würzburg